Ben Stacy Jerrik (Ed.)

Wuppertal-Oberbarmen Station

Ben Stacy Jerrik (Ed.)

Wuppertal-Oberbarmen Station

Elberfeld–Dortmund railway, Wuppertal-Oberbarmen–Solingen railway, Bergisch-Märkische Railway Company

Part Press

Contents

Articles

References

Wuppertal-Oberbarmen_station

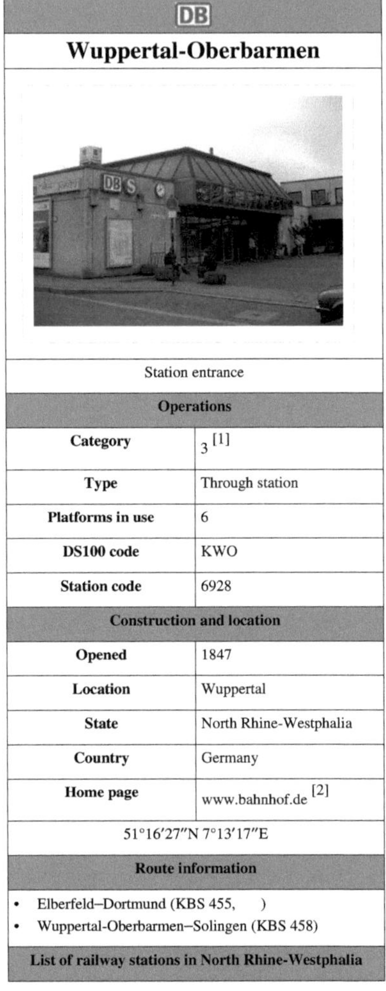

Wuppertal-Oberbarmen

Station entrance

Operations	
Category	3 [1]
Type	Through station
Platforms in use	6
DS100 code	KWO
Station code	6928
Construction and location	
Opened	1847
Location	Wuppertal
State	North Rhine-Westphalia
Country	Germany
Home page	www.bahnhof.de [2]
51°16′27″N 7°13′17″E	
Route information	
• Elberfeld–Dortmund (KBS 455,) • Wuppertal-Oberbarmen–Solingen (KBS 458)	
List of railway stations in North Rhine-Westphalia	

Wuppertal-Oberbarmen station is a station in the city of Wuppertal in the German state of North Rhine-Westphalia. It was long an important railway junction, connecting to four railway lines. The only remaining lines at the station are the Dortmund–Wuppertal main line and the branch line to Solingen.

History

January 1982

The first station building was opened along with the Elberfeld–Dortmund line under the name of *Barmen-Rittershausen* by the Bergisch-Märkische Railway Company on 9 October 1847. In 1930 it was renamed as *Wuppertal-Oberbarmen*.

In 1910, the tracks and Rosenau street were moved during the building of a depot at Wuppertal-Langerfeld. During the Second World War the station area and the station building were badly damaged. After a partial demolition by Deutsche Bundesbahn after the Second World War, the station was rebuilt in the 1980s during the establishment of S-Bahn line S8. Today there is a square-shaped commercial building with a newsagent, a bakery shop and a McDonald's branch.

In its heyday there was in addition to the Elberfeld–Dortmund through line, a triangular junction connected to the line to Opladen and Solingen, as well as a connecting line to the Düsseldorf-Derendorf–Dortmund Süd railway and the Wuppertal-Wichlinghausen–Hattingen line.

For a long time, Wuppertal-Oberbarmen was also an important freight terminal. The last freight tracks were removed in 2006, however, and a DIY store was built on the site.

Current operations

Long-distance passenger trains pass through Wuppertal-Oberbarmen without stopping. However, all regional trains running through Wuppertal stop. The Wupper-Express (RE 4), the Rhein-Münsterland-Express (RE 7) and the Maas-Wupper-Express (RE 13) stop at the station at hourly intervals. S-Bahn line S8 and Der Müngstener (RB 47) stop every twenty minutes on the local platforms and connect the station with Hagen, Mönchengladbach, Remscheid and Wuppertal Hauptbahnhof. One in three S-Bahn services terminate at Oberbarmen and do not run to or from Hagen.

Deutsche Bahn classifies the station as category 3.[1]

Schwebebahn eastern terminus at
Wuppertal-Oberbarmen station

Wuppertal-Oberbarmen is also a major connecting point between the railway and other public transport services. The Schwebebahn has its eastern terminus here, and there is a bus station, which is served by many of the lines of *Wuppertaler Stadtwerke* (Wuppertal's operator of public utilities and transport) and *Verkehrsgesellschaft Ennepe-Ruhr* (the transport company of Ennepe-Ruhr).

All passenger trains stopping at the station are operated by DB Regio NRW, except for the *Maas-Wupper-Express*, which is operated by Eurobahn under a 16-year concession with effect from the December 2009 timetable change.

Platforms

Today, there are three platforms with a total of six tracks. Regional trains stop on tracks 2 and 3; they are also used for non-stop operations by long-distance trains. Services on S-Bahn line S 8 and RB 47 stop on tracks 5 and 6; these tracks are the only ones with barrier-free access for the disabled. Barrier-free boarding is only possible on the S-Bahn line with its modern class 422 electric multiple units (since 2010). The class 628 diesel multiple units that serve the RB 47 route do not have barrier-free entry, although, under tenders called for the operation of the route in November 2009, the Verkehrsverbund Rhein-Ruhr requires that the operator of the route from 2013 use modern diesel multiple units that are accessible by the handicapped.

Interchanges

Service intervals below refer to peak hours from Mondays to Fridays.

Line	Line name	Route	Service interval	Platform track
	Wupper-Express	Aachen – Mönchengladbach – Düsseldorf – **Wuppertal** – Hagen – Dortmund	hourly	2/3
	Rhein-Münsterland-Express	Krefeld – Neuss – Cologne – Solingen – **Wuppertal** – Hagen – Hamm – Münster (Westf) – Rheine	hourly	2/3
	Maas-Wupper-Express	Venlo – Viersen – Mönchengladbach – Düsseldorf – **Wuppertal** – Hagen – Hamm (Westf)	hourly	2/3
	Der Müngstener	Solingen – Remscheid – **Wuppertal**	20 minutes	5/6
	Rhine-Ruhr S-Bahn	Hagen – Gevelsberg – **Wuppertal** – Düsseldorf – Neuss – Mönchengladbach	20 minutes	5/6

Preceding station	Deutsche Bahn	Following station
Wuppertal-Barmen *toward Aachen*	RE 4 *Wupper-Express*	Schwelm *toward Dortmund*
Wuppertal *toward Krefeld*	RE 7 *Rhein-Münsterland-Express*	Schwelm *toward Rheine*
Wuppertal-Barmen *toward Wuppertal*	RB 47 *Der Müngstener*	Wuppertal-Ronsdorf *toward Solingen*
Preceding station	**eurobahn**	**Following station**
Wuppertal-Barmen *toward Venlo*	RE 13 *Maas-Wupper-Express*	Schwelm *toward Hamm*
Preceding station	**Rhine-Ruhr S-Bahn**	**Following station**
Wuppertal-Barmen *toward Mönchengladbach*	S8	Wuppertal-Langerfeld *toward Hagen*

Notes

[1] "Station categories 2012" (http://www.deutschebahn.com/site/shared/de/dateianhaenge/infomaterial/sonstige/ bahnhofskategorieliste__db__station__service__2012.pdf) (in German) (PDF, 233.9 kb). Deutsche Bahn. . Retrieved 29 April 2012.

[2] http://www.bahnhof.de/site/bahnhoefe/en/bahnhofssuche__deutschland/bahnhofssuche/ bahnhofsdaten__filter__en,variant=details,recordId=6928.html

References

• WSW mobil Gmb, ÖPNV Systemmanagement (2009) (in German). *Wuppertal timetable 2009/10*. Wuppertal: ECO-Druck GmbH.

External links

- "Operations in KWO station area" (http://nrwbahnarchiv.bplaced.net/kln/KWO.htm) (in German). *NRW Rail Archive*. André Joost. Retrieved 4 September 2011.
- "Wuppertal-Oberbarmen station" (http://nrwbahnarchiv.bplaced.net/bf/8006719.htm) (in German). *NRW Rail Archive*. André Joost. Retrieved 4 September 2011.
- "Wuppertal-Oberbarmen station" (http://www.bahnen-wuppertal.de/html/bahnhof-oberbarmen.html) (in German). *Station portraits*. Bahnen Wuppertal. Retrieved 4 September 2011.

Elberfeld–Dortmund_railway

The **Elberfeld–Dortmund railway** is a major German railway. It is part of a major axis for long distance and regional rail services between Wuppertal and Cologne, and is served by Intercity Express, InterCity, Regional Express, Regionalbahn and S-Bahn trains.

This 56 km long line was the main line of the Bergisch-Märkische Railway Company. It was opened in 1849 and has been redeveloped several times since and is now fully electrified.

History

Since the Cologne-Minden Railway Company had decided to build its route via Duisburg rather than through the valley of the Wupper river, the *Bergisch-Märkische Railway Company* (German: *Bergisch-Märkische Eisenbahn-Gesellschaft*, BME) determined to build its own line through the Wupper valley, to create a link between the highly industrialised area of the Bergisches Land with the east, particularly to connect with the Märkische coal fields, near Dortmund. On 12 July 1844, it acquired a concession from the Prussian government for a rail link in the highly industrialised area of the Wupper valley and the Bergisch land. The line was opened from Döppersberg in Wuppertal to Schwelm on 9 October 1847. It was extended to Hagen and Dortmund on 20 December 1848.[1] [2]

On 9 March 1849, Düsseldorf-Elberfeld Railway Company's and the BME completed a line connecting the Elberfeld–Dortmund line with the Düsseldorf–Elberfeld line, which connected Wuppertal-Steinbeck to the Rhine at Düsseldorf and was completed in 1841.[1] This line ran through the new Elberfeld station (now called Wuppertal Hauptbahnhof) and the nearby Döppersberg station was closed.

Development of the Main Line

After the BME was nationalised, its main line between Düsseldorf and Dortmund were gradually upgraded. Between 1900 and 1915 additional tracks were built for local traffic with the existing tracks being reserved for through trains. The first section of track was opened in 1911 between Unterbarmen and Oberbarmen, followed by the section between Elberfeld and Vohwinkel opened on 10 April 1913. Two years later, the gap was closed between Elberfeld and Unterbarmen.[3]

Development for the S-Bahn

Following the establishment of the S-Bahn line S8 (Mönchengladbach–Hagen), the slow lines between Wuppertal-Vohwinkel and Wuppertal-Oberbarmen were incorporated into the S-Bahn line. On 29 May 1988 new sections of track were opened between Wuppertal-Oberbarmen and Linderhausen junction to the Schwelm–Witten line in the east and between Wuppertal-Vohwinkel and Dusseldorf Hbf in the west.[3]

Current situation

The line runs partly parallel with the tracks of the Wuppertal Northern Railway built by the Rhenish Railway Company, which ran through the countryside of the northern Wupper valley and is now largely abandoned.

The S-Bahn line S8 from Hagen to Dusseldorf and Mönchengladbach uses sections of both routes, which are connected by a short section of the largely abandoned Witten-Schwelm line. The S-Bahn line S9 connecting Dortmund, Witten and Hagen (where it connects with S8) runs entirely on the historic BME route.

Services

The line is also used every hour by Intercity Express line 10, connecting Cologne and Berlin via Hamm, Hanover, stopping at Wuppertal and Hagen. Additional InterCity trains also operate between Cologne and Hamm on IC lines 31 and 55 every 2 hours.

The section of the line between Hagen and Wuppertal is served hourly by Regional-Express line RE 4 (Wupper-Express) between Dortmund and Aachen via Dusseldorf, stopping at major stations. The line is also served hourly by line RE 7 (Rhein-Münsterland-Express) between Krefeld and Münster via Cologne and Hamm and by line RE 13 (Maas-Wupper-Express) between Venlo (Netherlands) and Hamm via Mönchengladbach. Various Regionalbahn services also operate on the line.

Notes

[1] "Line 2550: Aachen - Kassel" (http://nrwbahnarchiv.bplaced.net/strecken/2550.htm) (in German). *NRW Rail Archive*. André Joost. . Retrieved 28 October 2011.

[2] "Line 2801: Hagen-Witten–Dortmund" (http://nrwbahnarchiv.bplaced.net/strecken/2550.htm) (in German). *NRW Rail Archive*. André Joost. . Retrieved 28 October 2011.

[3] "Line L2525: Neuss-Schwelm–Linderhausen" (http://nrwbahnarchiv.bplaced.net/strecken/2550.htm) (in German). *NRW Rail Archive*. André Joost. . Retrieved 28 October 2011.

References

* Track data from *Eisenbahnatlas Deutschland (German railway atlas)*. Schweers + Wall. 2007. ISBN 978-3-89494-136-9.

Wuppertal-Oberbarmen–Solingen_railway

The **Wuppertal-Oberbarmen–Solingen railway** is a line in the Bergisches Land in the German state of North Rhine-Westphalia, which connects the three Bergisch cities of Wuppertal, Remscheid and Solingen. It is classified as a main line and is double track and non-electrified.

Today's route is made up largely of sections of three formerly independent routes built by the Bergisch-Märkische Railway Company (German: *Bergisch-Märkische Eisenbahn-Gesellschaft*, BME). The section between Remscheid and Solingen was built after the BME's nationalisation by the Prussian state railways.

History

The modern line between Solingen and Wuppertal includes several sections that originally formed parts of several independent lines:

- the Solingen–Wuppertal-Vohwinkel railway,
- the Wuppertal-Oberbarmen–Opladen railway,
- the Lennep–Hasten railway.

The section between Remscheid and Solingen was built after the BME's nationalisation by the Prussian state railways.

Ohligs Wald–Solingen (Weyersberg)

In 1867, the *Bergisch-Märkische Railway*, after many years of negotiations, began work on the construction of a 5.6-kilometre spur line to the city of Solingen (well known for its production of knives and scissors) from *Ohligs* station (then called *Ohligs Wald* and now called Solingen Hauptbahnhof). This line was completed on 25 September 1867.[1] In that year, the Gruiten–Köln-Deutz railway was completed.[2]

The line ended at a terminal station in Solingen Weyersberg, which was some distance from the city, but the line was primarily used for freight and passenger services were limited.

Weyersberg junction–Wuppertal-Vohwinkel

20 years later, on 12 February 1890, Solingen received a much better connection towards the Ruhr with the opening of the Solingen–Wuppertal-Vohwinkel line, known as the *Korkenzieherbahn* ("Corkscrew Railway"), to Vohwinkel.[3]

Passenger services were abandoned to Solingen-Weyersberg with the opening of the "Corkscrew Railway".[4] Trains ran from Weyersberg junction in a wide arc to the south around the city of Solingen, through the newly opened *Solingen Süd* station, which was significantly closer to the city of Solingen.

Solingen-Weyersberg station was now only a freight yard. It was closed along with the branch line from Weyersberg junction in 1912.

Wuppertal-Oberbarmen–Remscheid

On 1 September 1868, Remscheid was connected at Wuppertal-Oberbarmen (then called *Rittershausen*) with the Elberfeld–Dortmund railway, opened in 1849. The *Bergisch-Märkische Railway Company* opened the line from Rittershausen via Lennep to Remscheid as part of its Wuppertal-Oberbarmen–Opladen line. It gained the concession to build the last section of this line between Bergisch Born and Opladen only on 12 June 1872.[5] The section between Lennep and Remscheid was originally planned and built as a branch line of the Wuppertal-Oberbarmen–Opladen line and was also opened on 1 September 1868.[6]

A 2.4 km long branch from Remscheid Hauptbahnhof to Bliedinghausen was built in 1896, which has only ever been used for freight.[7]

Solingen–Remscheid

The remaining section between Remscheid and Solingen crossed very difficult topography and it was not completed until 1897 following the construction of several bridges.[1] A line was opened on 14 December 1893 from Solingen Süd Station for the transport of construction of material to Müngsten Bridge, which, when opened on 15 July 1897, completed the line to Remscheid.[1]

Already on 17 May 1897 a second track was opened between Ohligs and Solingen Süd and duplication was completed to Remscheid in 1907. Solingen Süd was expanded and renamed *Solingen Hauptbahnhof* in 1913.[1] In 1914, Remscheid station was renamed *Remscheid Hauptbahnhof*.

Further development

The line in Blombach valley

Both passenger and freight traffic on the line increased steadily. Remscheid-Lennep had become a major station, at the junction of three lines: the Wipper Valley Railway to Wipperfurth and Marienheide, the connection to Krebsoge station on the Wupper Valley Railway between Radevormwald and Wuppertal-Oberbarmen, and the Lennep–Opladen line. An engine depot was also built in Lennep for operations on the routes branching from it.

After the Second World War operations changed from steam to diesel haulage and in the 1970s freight traffic declined. Only a few companies on the line are now served by rail freight. There is still significant passenger traffic, which is now served with diesel multiple units.

Operations

A class 628 train in Remscheid-Güldenwerth station

The line is served along its entire length—and beyond it to Wuppertal Hauptbahnhof—by a Regionalbahn service, *Der Müngstener* (RB 47). The name refers to the meeting of the three cities of Solingen, Remscheid and Wuppertal near the former settlement of Müngsten at the line's crossing of the Wupper over Müngsten Bridge, the highest steel railway bridge in Germany.

On weekdays, trains run during the day at 20-minute intervals, in the evenings and on weekends, every 30 minutes. Not all the trains run over the entire route; some trips (from both directions) terminate in Remscheid.

Deutsche Bahn has operated the service with class 628 diesel multiple units since 1994. There are many cancellations due to the vehicles' lack of power and the hilly route, especially in the autumn. Following the tendering of services on the line from November 2009 to July 2010, a new operator, Abellio Rail NRW, will take over operations from 2013 using more powerful (LINT 41) vehicles.

Modernisation

The stations on the route have been upgraded over several years. This is especially evident in Solingen, where the former Hauptbahnhof (central station) was closed in May 2006. A replacement station was opened at the same time at *Solingen Grünewald*, not far away from the old Hauptbahnhof and much better connected to the city and buses. At the timetable change in December 2006, a second new station was opened at *Solingen Mitte*.

The name of Solingen Hauptbahnhof has since 2006 been applied to the former Solingen-Ohligs station, which is the only stop for Regional-Express, Intercity and Intercity-Express in the city of Solingen.

Solingen Mitte station

Structures

Bridges

Due to the mountainous landscape, the railway required the building of numerous structures such as bridges and tunnels. The highest railway bridge at 107 m, is Müngsten Bridge, which crosses the valley of the Wupper between Remscheid and Solingen. Every autumn the so-called *Brückenfest* ("bridge festival") is held, during which special trains are also operated on the line. The bridge was inaugurated in 1897 and is the largest engineering structure on the route.

There are a number of other bridges, especially on the section between Solingen and Remscheid, Only a few hundred metres away there is Windfeln Bridge, which is considered the "little sister" of Müngsten Bridge.

A train on Müngsten Bridge

Tunnels

Another notable structure is the Rauenthal Tunnel in Wuppertal, through which the railway runs between Oberbarmen and Ronsdorf under a mountain range and the residential area located on it. Two parallel tunnel tubes were cut through the mountain for the double-track line.

Directly next to the southern portal of the two tunnels there is a third, which was part of the former Langerfeld Tunnel, connecting the Wuppertal-Rauenthal freight yard with the container yard of Wuppertal-Langerfeld. This tunnel is now closed.

As the eastern tube of Rauenthal Tunnel has been blocked since 2005 for safety reasons, trains only use the western tube. South of the tunnel, Rauenthal station ends in a deep cutting.

The southern portal of the Rauenthal Tunnel, on the right is the Langerfeld Tunnel

Notes

[1] "Line 2675: Solingen Hbf ↔ Remscheid Hbf" (http://nrwbahnarchiv.bplaced.net/strecken/2675.htm) (in German). *NRW Rail Archive*. André Joost. . Retrieved 26 October 2011.

[2] "Line 2730: Gruiten - Neurather Ring" (http://home.arcor.de/nrwbahnarchiv/strecken/2730.htm) (in German). *NRW Rail Archive*. André Joost. . Retrieved 25 October 2011.

[3] "Line 2734: Solingen - Wuppertal-Vohwinkel" (http://nrwbahnarchiv.bplaced.net/strecken/2734.htm) (in German). *NRW Rail Archive*. André Joost. . Retrieved 26 October 2011.

[4] "Solingen (Weyersberg) station operations" (http://nrwbahnarchiv.bplaced.net/ex/xSWB.htm) (in German). *NRW Rail Archive*. André Joost. . Retrieved 26 October 2011.

[5] "Line 2700: Wuppertal-Oberbarmen ↔ Remscheid-Lennep (↔ Opladen)" (http://nrwbahnarchiv.bplaced.net/strecken/2700.htm) (in German). *NRW Rail Archive*. André Joost. . Retrieved 26 October 2011.

[6] "Line 2705: Remscheid-Lennep ↔ Remscheid Hbf (↔ Remscheid-Hasten)" (http://nrwbahnarchiv.bplaced.net/strecken/2705.htm) (in German). *NRW Rail Archive*. André Joost. . Retrieved 26 October 2011.

[7] "Line 2706: Remscheid Hbf ↔ Remscheid-Bliedinghausen" (http://nrwbahnarchiv.bplaced.net/strecken/2706.htm) (in German). *NRW Rail Archive*. André Joost. . Retrieved 26 October 2011.

External links

- Lothar Brill. "Line 2675" (http://www.eisenbahn-tunnelportale.de/lb/inhalt/tunnelportale/2675.html) (in German). *Photographs of tunnel portals*. www.eisenbahn-tunnelportale.de. Retrieved 26 October 2011.
- Lothar Brill. "Line 2700" (http://www.eisenbahn-tunnelportale.de/lb/inhalt/tunnelportale/2700.html) (in German). *Photographs of tunnel portals*. www.eisenbahn-tunnelportale.de. Retrieved 26 October 2011.

Bergisch-Märkische_Railway_Company

The **Bergisch-Märkische Railway Company** (German: *Bergisch-Märkische Eisenbahn-Gesellschaft*, BME) was a Germany railway company that together with the Cologne-Minden Railway (*Cöln-Mindener Eisenbahn-Gesellschaft*, CME) and the Rhenish Railway Company (*Rheinische Eisenbahn-Gesellschaft*, RhE) was one of the three (nominally) private railway companies that in the mid-19th century built the first railways in the Ruhr and large parts of today's North Rhine-Westphalia. Its name refers to Bergisches Land and the County of Mark.

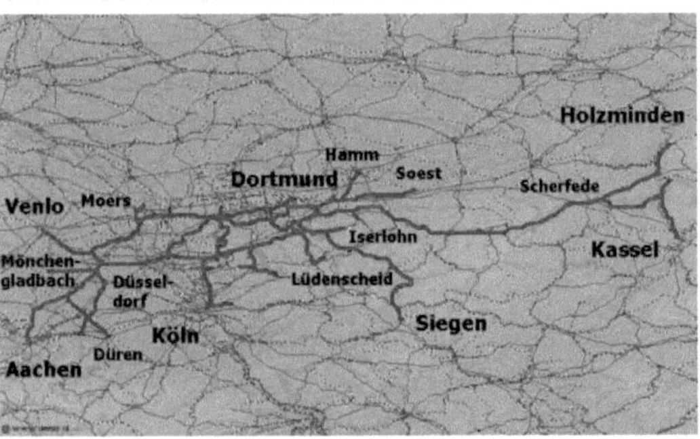

Network of the *Bergisch-Märkische Railway Company* shortly before its nationalisation in 1882

History

Foundation

The Bergisch-Märkische Railway Company was founded on 18 October 1843 in Elberfeld (Today Wuppertal). Since the Cologne-Minden Railway Company had decided to build its route via Duisburg rather than through the valley of the Wupper river, the *Bergisch-Märkische Railway Company* (German: *Bergisch-Märkische Eisenbahn-Gesellschaft*, BME) determined to build its own line through the Wupper valley, to create a link between the highly industrialised area of the Bergisches Land with the east, particularly to connect with the Märkische coal fields, near Dortmund. The required concession for the railway was Granted by the Prussian government on 12 July 1844. A link to the Rhine in the west had already been completed in 1841 by the Düsseldorf-Elberfeld Railway Company, which had been founded in 1837.

Trunk routes

Its original, 56 km long main line ran from Elberfeld to Dortmund via Barmen (since 1929 part of Wuppertal), Schwelm, Hagen, Wetter and Witten and was completed in 1849. In the following years the company built other main and branch lines in the Ruhr along the Hellweg an ancient highway and the Ruhr and Rhine rivers. In 1862 it opened a profitable east-west trunk line between Dortmund and Witten through Bochum-Langendreer, Essen, Mülheim an der Ruhr to Duisburg. The development of the Ruhr valley was largely a result of the opening of the BME's trunk line.

Line at Schwelm in 1861

The company's development was characterised by the acquisitions of many smaller railway companies to round out its network. However, its energetic board of directors and its chairman Daniel von der Heydt (1802–1874, later a member of the Prussian House of Lords), despite years of effort, were not able to take over the Prussian government-owned Royal Westphalian Railway Company. Such a takeover would have allowed the BME to develop a connection via Hamm to a German seaport via Rheine.

Expansion

Major expansion began in 1859 with the construction of the 106 km long Ruhr–Sieg from Hagen to Siegen and its mines. The line opened on 6 August 1861 and cost 12.9 million thalers. In 1858 it started to build its Witten–Duisburg trunk line through the Ruhr. The first section was opened between Duisburg and Hochfeld for freight trains only on 19 August 1859. The 52 km line from Bochum-Langendreer to Steele, Essen and Mülheim an der Ruhr, with connections to various coal mines, was completed on 1 May 1862. At Steele it also connected with the northern end of the Steele–Vohwinkel railway, which had been rebuilt in 1847 from the Prince William Railway (opened as the first horse-powered railway in Germany in 1831) and acquired by the BME in 1854 for 1.3 million thalers.

Logically, then its next step in 1866 was to cross the Rhine via the Ruhrort–Homberg train ferry with the goal of connecting with Belgium and Netherlands through the purchase of the Aachen-Düsseldorf-Ruhrort Railway Company's lines for seven million thalers. In 1870, it completed the Hamm railway bridge bridge across the Rhine in Düsseldorf-Hamm and opened the line from Dusseldorf to Neuss. This created a second connection between its networks on the east and west banks of the Rhine.

In addition the construction of several smaller routes followed up to 1876, an extension in an easterly direction, the Upper Ruhr Valley Railway to Arnsberg, Bestwig, Brilon-Wald and Warburg and Holzminden on the Weser river. Here it connected with the line to Kassel of the Frederick William Northern Railway Company, which it took over on 17 April 1868, with its 130-kilometer line from Gerstungen via Bebra and Kassel to Bad Karlshafen for eight million thalers. After 1870 the network was extended on the west bank of the Rhine with the 66 km long line from Rheydt-Odenkirchen to Aue and Düren. During the nationalisation of the company in 1880 the company took over the 78 km railway network of the Dutch-Westphalian Railway Company from Gelsenkirchen-Bismarck to Dorsten, Borken to Winterswijk in the Netherlands, with a branch from Borken to Bocholt.

Opening and acquisition of lines

Dates	Lines	Route
1847–1849	Elberfeld–Dortmund	**Trunkline**: Elberfeld–Oberbarmen–Schwelm–Milspe-Hagen–Witten–Dortmund
1855	Dortmund–Soest	Dortmund–Hörde–Unna–Soest
1857	Düsseldorf–Elberfeld	*acquisition of the Dusseldorf-Elberfeld Railway Company*
1859–1861	Ruhr–Sieg	Hagen–Letmathe–Finnentrop–Kreuztal–Siegen
1859–1862	Witten/Dortmund–Oberhausen/Duisburg	Witten/Dortmund–Langendreer–Essen–Mülheim (Ruhr)–Mülheim-Styrum–Oberhausen/Duisburg
1863	Wuppertal-Vohwinkel–Essen-Überruhr	*acquisition of the Prince William Railway Company*
1864	Aachen–Mönchengladbach	*acquisition of the Aachen-Düsseldorf Railway*
	Mönchengladbach–Düsseldorf	
	Duisburg-Ruhrort–Mönchengladbach	
1866	Viersen–Venlo	Viersen–Kaldenkirchen–Venlo
1866/7	Hagen–Hamm	Unna–Hamm (1866), Hagen–Herdecke (1867)
1867/8	Gruiten–Köln-Deutz	(Elberfeld–)Gruiten–Solingen–Opladen–Köln-Deutz(–Cologne)
1868	Friedrich-Wilhelms Northern Railway	*acquisition of the Friedrich-Wilhelms Northern Railway Company*
1868-81	Wuppertal-Oberbarmen–Opladen/Remscheid	Oberbarmen–Lennep–Remscheid (1868), Lennep–Wermelskirchen (1876), Wermelskirchen–Opladen (1881)
1870	Hamm railway bridge	Neuss–Düsseldorf
1870–73	Schwerte–Warburg	Schwerte–Arnsberg (1870) Arnsberg–Meschede (1871) Meschede–Bestwig (1872) Bestwig–Warburg (1873)
1872	Düsseldorf–Schwerte	Düsseldorf–(Essen-)Kettwig–(Essen-)Kupferdreh–Herdecke–Schwerte
1873	Hochneukirch–Stolberg	Rheydt-Odenkirchen–Jülich–Düren, Jülich–Weisweiler–Aue
1876	Mülheim-Styrum–Essen-Kettwig	(Mülheim-)Styrum–(Essen-)Kettwig
1876	Scherfede–Holzminden	Scherfede–Beverungen–Holzminden
1877	Essen-Werden–Essen	(Essen-)Werden–Essen
1879	Iron Rhine	Rheydt–Dalheim–D/NL border at Vlodrop
1880	Gelsenkirchen-Bismarck–Winterswijk	*acquisition of Dutch-Westphalian Railway Company*

Nationalisation

The act for the nationalisation of the Bergisch-Märkische Railway Company was promulgated on 28 March 1882. At that time, the Prussian government held 64 percent of the share capital of the Company. The Prussian state railways's *Royal directorate of railways at Elberfeld* (German: *Königliche Eisenbahn-Direction zu Elberfeld*) took over its management with effect from 1 January 1882.

At its nationalisation the company had 768 locomotives and 21,607 wagons. Its rail network was 1,336 km long, including 720 km of double track railway. The purchase price was financed by government bonds worth 633,847,500 marks. The company was dissolved on 1 January 1886.

References

- (in German) *Das Bergisch-Märkische Eisenbahn-Unternehmen in seiner Entwicklung während der ersten 25 Jahre des Betriebe.* Elberfeld: Bergisch-Märkische Railway Company. 1875.
- (in German) Annual Reports of the Bergisch-Märkisch Railway Company
- Regierungsassessor Waldeck (1910). "Die Entwicklung der Bergisch-Märkischen Eisenbahnen (The development of Bergisch-Märkisch Railways)" (in German). *Archiv für Eisenbahnwesen (Archive ofRailway Engineering).* **3-5**. Berlin.
- (in German) *Die Deutschen Eisenbahnen in ihrer Entwicklung 1835–1935 (The German railways in its development 1835-1935).* Berlin: Deutsche Reichsbahn. 1935.
- Rolf, Ostendorf (1979) (in German). *Eisenbahn-Knotenpunkt Ruhrgebiet, Die Entwicklungsgeschichte der Revierbahnen seit 1838 (Ruhr rail hub, the history of the area's line since 1838.* Stuttgart.
- Klee, Wolfgang; Scheingraber, Günther (1992). "Preußische Eisenbahngeschichte (Prussian Railway History, Part 1: 1838-1870)" (in German). *Preußen-Report (Prussian report).* **1.1**. Fürstenfeldbruck.
- Saling, A. (1869) (in German). *Die norddeutschen Börsen-Papiere 2. Jg. 1868-1869 (The North German stock exchange papers).* Berlin.

Wupper-Express

The **Wupper-Express** (RE 4) is a Regional-Express service in the German state of North Rhine-Westphalia (NRW) running from Aachen via Mönchengladbach, Düsseldorf, Wuppertal, Hagen to Dortmund. The service is operated every hour by DB Regio NRW. It is the third most widely used Regional-Express line in the area administered by the Verkehrsverbund Rhein-Ruhr with approximately 24,000 passengers a day.[1]

History

Today's RE 4 is the successor to the former *StädteExpress* line SE from Aachen to Hagen and Iserlohn. Later, the end point was moved to Hamm and after the abolition of InterRegio services it was extended to Munster. Under the second stage of North Rhine-Westphalia's integrated timetable (ITF 2), introduced in December 2002, it was replaced by the Maas-Wupper-Express (RE 13) and the Ems-Börde-Bahn (RB 89) services between Hagen and Munster and the Wupper-Express has since then run to Dortmund with a stop in Witten.

Route

The Wupper-Express runs successively over the Aachen–Mönchengladbach, the Mönchengladbach–Düsseldorf and the Düsseldorf–Elberfeld lines. The service then follows the Elberfeld–Dortmund railway as far as Witten station, from where it uses the tracks of the Witten/Dortmund–Oberhausen/Duisburg railway and the Oberstraße Tunnel on its way to Dortmund station. At night, the RE 4 operates to Düsseldorf Airport Terminal station.

The Wupper-Express runs parallel to Rhine-Ruhr S-Bahn lines on large sections of track and it has some of the character of a fast S-Bahn service and is perceived by passengers accordingly.

Rollingstock

The Wupper-Express uses class 111 locomotives and non-air conditioned double-deck coaches. Additional peak hour services operate between Düsseldorf and Aachen with class 110 and 111 locomotives, operated exclusively with refurbished Silberling carriages.

Wupper-Express near
Baal

Preparation of the
train in Aachen

Train running label

Leaving
Aachen
Hbf

Wupper-Express on track 3 in
Aachen Hbf

Notes

[1] "Qualitätsbericht SPNV Im Verkehrsverbund Rhein-Ruhr für 2010" (http://vrr.de/imperia/md/content/publikationensonstige/qualitaetsbericht.pdf) (in German) (PDF). Verkehrsverbund Rhein-Ruhr. February 2011. . Retrieved 7 September 2011.

See also

- List of regional rail lines in North Rhine-Westphalia
- List of scheduled railway routes in Germany

External links

- "Wupper-Express" (http://nrwbahnarchiv.bplaced.net/linien/RE4.htm) (in German). *NRW rail archive*. André Joost. Retrieved 7 September 2011.

Deutsche_Bundesbahn

The **Deutsche Bundesbahn** or **DB** (German Federal Railway) was formed as the state railway of the newly established Federal Republic of Germany (FRG) on September 7, 1949 as a successor of the Deutsche Reichsbahn-Gesellschaft **(DRG)**. The DB remained the state railway of West Germany until after German reunification, when it was merged with the former East German Deutsche Reichsbahn **(DR)** to form Deutsche Bahn AG, which came into existence on 1 January 1994.

Logo of the DB, "der Keks" (the biscuit), used 1949-1994

Background

After World War II, each of the military governments of the Allied Occupation Zones in Germany was *de facto* in charge of the German railways in their respective territories. On 10 October 1946, the railways in the British and American occupation zones formed the *Deutsche Reichsbahn im Vereinigten Wirtschaftsgebiet* (German Imperial Railway in the united economic area), while on 25 June 1947, the provinces under French occupation formed the Südwestdeutsche Eisenbahn. With the formation of the FRG these successor organizations of the DRG were reunited, a situation codified by the **Federal Railways Law** *(Bundesbahngesetz)* that was ratified on 13 December 1951. The railways in the Saarland joined on 1 January 1957.

Bahndienstfernschreiben

B VON BONN   NR 13= 06.9.49 20.30 ==
AN HAUPTVERWALTUNG DER DEUTSCHEN REICHSBAHN IM
VEREINIGTEN WIRTSCHAFTSGEBIET , OFFENBACH / MAIN=
MIT WIRKUNG VOM 7. SEPTEMBER 1949 WIRD DIE BEZEICHNUNG
DEUTSCHE REICHSBAHN IM VEREINIGTEN WIRTSCHAFTSGEBIET IN
DEUTSCHE BUNDESBAHN   GEAENDERT./DER NAME DEUTSCHE
REICHSBAHN IST AUCH IN ALLEN BEHOERDENBEZEICHNU GEN DURCH
DEUTSCHE BUNDESBAHN   IN DER MIT DEN SUEDWESTDEUTSCHEN
EISENBAHNEN VERABREDETEN FORM ZU ERSETZEN/ DIE
BEZEICHNUNG DER EISENBAHNBEHOERDEN IN DER FRANZOESISCHEN
ZONE WIRD DURCH DIE DAFUER ZUSTAENDIGEN STELLEN GERGELT
DIE HIERAUS SICH ERGEBENDE AENDERUNG DER
AMTSBEZEICHNUNGEN DER REICHSBAHNBEAMTEN IM VEREINIGTEN

WIRTSCHAFTSGEBIET WIRD BESONDERS VERFUEGT. ===
DER DIREKTOR DER VERWALTUNG FUER VERKEHR

GEZ. DR. ING. F R O H N E ===

Telegram announcing the formation of the Deutsche Bundesbahn

Organisation

The DB was a state-owned company that, with few local exceptions, exercised a monopoly concerning rail transport throughout West Germany. The DB was placed under the control of the *Bundesverkehrsministerium* (Federal Transport Ministry). With its headquarters in Frankfurt, in 1985 the DB was the third-largest employer in the FRG, with a strength of 322,383 employees. A special transit police *(Bahnpolizei)* provided security. The catering needs of the DB were supplied by the "Deutsche Schlafwagen- und Speisewagengesellschaft" **(DSG)**, later "Deutsche Service-Gesellschaft der Bahn", as the former DRG caterer Mitropa was situated in East Germany and serviced the Deutsche Reichsbahn in the GDR.

As West Berlin lay surrounded by the GDR, local and long-distance railway services in the divided city were provided exclusively by the DR, although the DB operated a ticket office in the Hardenbergstraße near the main West Berlin passenger station Zoologischer Garten.

1949-1970

The immediate tasks in the early years after the end of World War II involved the reconstruction of the heavily damaged infrastructure and the replenishment of locomotives and rolling stock. Contrary to the Deutsche Reichsbahn in the GDR, the DB was not subject to reparations and benefited from the influx of capital through the Marshall Plan. During the early years, new steam engines were constructed and placed into service. The last new steam locomotive type was the Class 10, which entered service in 1957. Only two units of class 10 were built. In 1959 DB took the last steam locomotive delivery when the last of the class 23 locomotives was delivered. Soon, with increase in mass motorization, the railway started to lose passenger volume. As a result rail buses were introduced on some lines, while other smaller volume lines were closed. Main lines became increasingly electrified. The later years of this epoch saw a decrease and eventual phasing out of steam engines, with the last one to cease regular service in 1977. Traction was provided increasingly by diesel and electric engines. With increased use of diesel and electric locomotives, progress was made in decreasing travel time for passengers. New types of passenger trains were introduced such as the Trans Europ Express and the InterCity.

Transport of goods also had to compete with the ever increasing competition from trucks. Furthermore, traditional services such as coal and iron ore shipments declined with the changes in the overall economy.

Trans Europ Express locomotive in West Germany, c. 1961.

V200 number 010 pulling passenger train in West Germany, c. 1961.

1970-1993

After the introduction of the TGV in France, the ICE system of high speed passenger trains was developed. Significant stretches of new high speed track, like the Hanover-Würzburg high-speed rail line, had to be laid or upgraded. Other characteristics of this epoch are the introduction of computer systems and the steps taken towards an integrated system of European railways. Externally, rolling stock displayed more colourful and varied livery schemes.

The two German states were reunified in October 1990 with both DB and DR now being special funds of the German Federal government. Article 26 of the Unification Treaty (Einigungsvertrag) stipulated the DR to be merged with DB at the earliest opportunity. The DB, in the interim, initiated new coordinations in businesses with the DR, started IC and ICE services into Berlin, and extended IC and ICE services to major cities in eastern Germany. Administratively, on June 1, 1992 the DB and DR formed a joint board of directors which governed both entities. However, the merger was delayed over the structure of merged railway due to concerns by German politicians on the ever-increasing annual operating deficits incurred by the DB and DR. After several years of delays, the Bundesverkehrsministerium proposed a comprehensive reform of the German railway system *(Bahnreform)* which was approved by the Bundestag in 1993 and went into effect on 1 January 1994. At the heart of the reform package was a) the merger of the DB and the DR and b) the change of the form of the enterprise into a stock corporation. Nevertheless, the Deutsche Bahn AG was still publicly owned.

Presidents of the DB

- Heinz Maria Oeftering, 13 May 1957 - 12 May 1972
- Wolfgang Vaerst, 13 May 1972 – 12 May 1982
- Reiner Gohlke, 13 May 1982 – 18 June 1990
- Heinz Dürr, 1 January 1991 - 31 December 1993 (thereafter the head of newly formed Deutsche Bahn AG).

See also

- Locomotive classification of the Deutsche Bundesbahn
- List of DB locomotives and railbuses
- Deutsche Bahn

External links

- Pictures [1]
- DB:1949-1970 (in German) [2]
- DB:1968-1990 (in German) [3]
- Some historical background (German) [4]

References

[1] http://www.bics.be.schule.de/son/verkehr/eisenbah/index.htm
[2] http://epoche-3.de/index.html
[3] http://bundesbahnzeit.de
[4] http://schorch.ch/bahn/texte/02.htm

Rhine-Ruhr_S-Bahn

S-Bahn Rhein-Ruhr S-Bahn Rhein-Sieg / S-Bahn Köln	
Background	
Locale	Rhine-Ruhr, North Rhine-Westphalia, Germany
Transit type	Commuter rail
Number of lines	13
Number of stations	124
Annual ridership	130 million Düsseldorf/Rhine-Ruhr: 98 million[1] Cologne: 32 million[2]
Headquarters	Düsseldorf, Germany
Website	www.s-bahn-rhein-ruhr.de [3] www.s-bahn-koeln.de [4]
Operation	
Began operation	1967
Operator(s)	DB DB Regio NRW, Regiobahn (S28)
Headway	20 min.
Technical	
System length	676 km (**unknown operator: u'strong'** mi)

The **Rhine-Ruhr S-Bahn** (German: *S-Bahn Rhein-Ruhr*) is a polycentric S-Bahn network covering the Rhine-Ruhr Metropolitan Region in the German federal state of North Rhine-Westphalia. This includes most of the Ruhr (and cities such as Dortmund, Duisburg and Essen), the Berg cities of Wuppertal and Solingen and parts of the Rhineland (with cities such as Cologne and Düsseldorf). The easternmost city within the S-Bahn Rhine-Ruhr network is Unna, the westermost city served is Mönchengladbach.

The S-Bahn network was established in 1967 with a line connecting Ratingen Ost to Düsseldorf-Garath and nowadays consists of 13 lines.

Branding of S-Bahn Rhein-Ruhr

It is operating in the area of the Verkehrsverbund Rhein-Ruhr and Verkehrsverbund Rhein-Sieg tariff associations, touching the areas of Aachener Verkehrsverbund at Düren and Verkehrsgemeinschaft Ruhr-Lippe at Unna.

History

X-Wagen coaches of S-Bahn Rhein-Ruhr at Cologne central station

The predecessor of the S-Bahn was the so-called *Bezirksschnellverkehr* between the cities of Düsseldorf and Essen, which consisted of steam-powered push-pull trains, mainly hauled by Class 78 and Class 65 engines.

The first S-Bahn lines were operated using Silberling cars and Class 141 locomotives, however these were not suited for operations on an urban network and were soon replaced by Class 420 electric multiple units. In the mid-1970s, the Class 420 was decided to be unsuitable for the network as well, mainly due to being uncomfortable and lacking a lavatory, since one could travel rather long distances on the Rhine-Ruhr network, which wasn't the case on the Munich S-Bahn for which the class 420 were originally designed. At first an improved version of the 420, the *Class 422*, was discussed, but in 1978 the Deutsche Bundesbahn commissioned a batch of coaches from Duewag and MBB, called the *x-Wagen* (the x-car) after its classification code *Bx*. In late 1978, the first prototypes (the 2nd class cars of type Bx 794.0 and the cab car Bxf 796.0) were handed over to the DB, the 1st/2nd class cars ABx 791.0 following in early 1979. The prototypes

Network map

were successful, and so from 1981 to 1994 several series were commissioned, first to be hauled by the Class 111 engines, but after the German reunification the surplus Reichsbahn engines of Class 143 replaced the 111s on the S-Bahn network.

Rolling stock today

On much of the network, Class 143 locomotives are used along with the specially developed *Bx* (second class) and *ABx* (second and first class) cars and cab cars (*Bxf*). The Class 420 electric multiple units previously belonging to the Munich, Stuttgart and Frankfurt networks which had operated the S7 and S9 services were finally retired at the beginning of 2009, and have been replaced with the new DBAG Class 422, while Class 423 EMUs can be found on the S11, S12 and S13 lines. The S28 is not operated by DB Regio NRW, but by the Regiobahn GmbH, which uses Bombardier TALENT DMUs on the line.

Lines

The region's lines were mainly built by three railway companies (the Cologne-Minden Railway Company, the Bergisch-Märkische Railway Company and the Rhenish Railway Company), giving the Rhine-Ruhr S-Bahn a variety of lines to use for its routes. This means that the S-Bahn lines use up to five different railways to run over.

Line	Route	Railways used	Length	Opening date of first section[5]	First section[5]
	Dortmund – Bochum – Essen – Mülheim (Ruhr) – Duisburg – Düsseldorf Airport – Düsseldorf – Hilden – Solingen	Dortmund–Duisburg, Duisburg–Düsseldorf, Düsseldorf–Solingen	97 km	26.05.1974	Bochum – DU-Großenbaum
	Dortmund – Dortmund-Dorstfeld – Dortmund-Mengede – Herne – (Gelsenkirchen – (Oberhausen – Duisburg) or Essen) or Recklinghausen	Dortmund–Dortmund-Dorstfeld, Dortmund-Dorstfeld–Dortmund-Mengede, Dortmund-Mengede–Herne/Gelsenkirchen/Duisburg, and part of Gelsenkirchen–Essen or Herne–Recklinghausen	58 / 42 / 33 km	02.06.1991	Dortmund – Duisburg
S 3	Oberhausen – Mülheim (Ruhr) – Essen – Essen-Steele – Hattingen (Ruhr) Mitte	Oberhausen–Essen-Steele Ost, Essen-Steele Ost–Bochum-Dahlhausen, Bochum-Dahlhausen–Hattingen (Ruhr) Mitte	33 km	26.05.1974	Oberhausen – Hattingen (Ruhr)
	Dortmund-Lütgendortmund – Dortmund–Dorstfeld – Unna-Königsborn – Unna	Dortmund-Lütgendortmund–*Dortmund Süd*, *Dortmund Süd*–Unna-Königsborn, Unna-Königsborn–Unna	30 km	03.06.1984	DO-Germania – Unna
	Dortmund – Witten – Wetter (Ruhr) – Hagen	Dortmund–Hagen	31 km	29.05.1994	*Whole length*
	Essen – Ratingen Ost – Düsseldorf – Langenfeld (Rheinl) – Cologne – Cologne-Nippes	Essen–Essen-Werden, Essen-Werden–Düsseldorf, Düsseldorf–Cologne, Cologne–Köln-Nippes	78 km	28.09.1967	Ratingen Ost – D-Garath
	Hagen – Wuppertal – Wuppertal-Vohwinkel – Düsseldorf – Neuss – Mönchengladbach	Hagen-Schwelm, Schwelm–Wuppertal, Wuppertal–Düsseldorf, Düsseldorf–Mönchengladbach	82 km	29.05.1988	*Whole length*
	Haltern am See – Gladbeck West – Bottrop – Essen – Essen-Steele – Velbert-Langenberg – Wuppertal-Vohwinkel – Wuppertal	Haltern–Essen-Dellwig Ost, Essen-Dellwig Ost–Essen West, Essen West–Essen-Steele, Essen-Steele–Wuppertal-Vohwinkel, Wuppertal-Vohwinkel–Wuppertal	90 km	24.05.1998	Haltern – Essen-Steele
	Düsseldorf Airport Terminal – Düsseldorf – Neuss – Cologne-Nippes – Cologne – Bergisch Gladbach	Düsseldorf Airport Terminal–Düsseldorf-Unterrath railway, Düsseldorf-Unterrath–Düsseldorf, Neuss–Cologne, Cologne–Köln-Mülheim, Cologne-Mülheim–Bergisch Gladbach	74 km	01.06.1975	K-Chorweiler – Berg. Gladbach
	Düren – Horrem – Cologne – Troisdorf – Siegburg/Bonn – Au (Sieg)	Düren–Cologne, Cologne–Au Sieg	105 km	02.06.1991	Köln-Nippes – Au (Sieg)
	Horrem – Cologne – Cologne/Bonn Airport – Troisdorf	Horrem–Cologne, Cologne–Troisdorf incl. Cologne Airport loop	45 km	15.12.2002	Düren – Cologne-Deutz
	Mettmann Stadtwald – Düsseldorf – Neuss – Kaarster See	Mettmann Stadtwald–Düsseldorf, Düsseldorf–Neuss, Neuss–Kaarster See	34 km	26.09.1999	*Whole length*

Wuppertal-Vohwinkel – Düsseldorf – Langenfeld (Rheinl)	Wuppertal–Düsseldorf, Düsseldorf–Langenfeld	39 km	13.12.2009	*Whole length*

Kursbuchstrecken 450.x (x is equivalent to the number of the line), as of 13 December 2009.

See also

- List of rapid transit systems

References

[1] Press note (http://www.deutschebahn.com/site/bahn/de/presse/presseinformationen/nrw/nrw20110128.html) Deutsche Bahn, 28. January 2011

[2] Facts and figures (http://www.s-bahn-koeln.de/regional/view/regionen/nrw/info/s-bahn_koeln_unternehmen.shtml) S-Bahn Köln

[3] http://www.s-bahn-rhein-ruhr.de/

[4] http://www.s-bahn-koeln.de/

[5] "S-Bahn Rhein-Ruhr-Sieg - Geschichte" (http://www.marco-wegener.de/s-bahn/index.htm) (in German). www.indusi.de. . Retrieved 25 August 2011.

External links

- Rheinbahn (Transportation Company of Düsseldorf) (http://www.rheinbahn.de)

- Official Website of Rhine-Ruhr S-Bahn (only German) (http://www.bahn.de/regional/view/regionen/nrw/info/a_s_bahn_rhein_ruhr.shtml)

- www.marco-wegener.de - Information and History of Rhine-Ruhr-Sieg S-Bahn (German) (http://www.marco-wegener.de/s-bahn/index.htm)

Rhein-Münsterland-Express

The **Rhein-Münsterland-Express** (RE 7) is a Regional-Express service in the German state of North Rhine-Westphalia (NRW). The hourly service initially runs to the south east from Krefeld via Neuss to Cologne and then turns to run to the northeast via Solingen, Wuppertal, Hagen to Münster. Every two hours it continues to Rheine.

History

Today's RE 7 is the successor to the former *StädteExpress* line SE from Aachen via Cologne and Wuppertal to Munster.

From 1998, under the original version of North Rhine-Westphalia's integrated timetable (ITF 1), the service ran between Düren and Munster. With the introduction of ITF 2 in December 2002, the line was extended at both ends to Aachen and to Rheine. Since the Rhein-Sieg-Express (RE 9) often ran late under the new timetable, in June 2003, the RE 7 exchanged its section on the left (west) bank of the Rhine with the RE 9's left bank route and has since then run to Krefeld. This eliminates a level crossing of rail tracks in Cologne and avoids conflicts between the two services. The short turn around times available for the RE 7 in Krefeld then led to significant delays on the RE 7.

To stabilise the timetable, in 2006 locomotives on the line were changed from class 111 to class 112 and the Rhein-Münsterland-Express since then has runs every two hours to Münster. These changes eventually led to normal on-time running.[1] The Zweckverband SPNV Münsterland (Münsterland rail transport association) is planning, however, to extend hourly services to Rheine. As part of the second stage of modernisation at Rheine station, one of the two disused platforms are to be reactivated, which would make available six platform tracks instead of four. The train could therefore operate without platform conflicts with InterCity services on the Emsland line. In return, Regionalbahn RB 68 services would be discontinued.

The line is operated every hour by DB Regio NRW. Long sections of RE 7 runs parallel to Rhine-Ruhr S-Bahn lines and it has some of the character of a fast S-Bahn service and is perceived by the passengers accordingly.

Route

The Rhein-Münsterland-Express runs on the following railway lines:

1. Rheine–Münster line (full length)
2. Münster–Hamm line (full length)
3. Hamm–Hagen line (full length)
4. Dortmund–Wuppertal line (from Hagen Hauptbahnhof)
5. Wuppertal-Düsseldorf line (to Linden/Gruiten junction)
6. Gruiten–Köln-Deutz line (to Köln-Mülheim)
7. Duisburg– Köln-Deutz line (from Cologne-Mülheim)
8. Hohenzollern Bridge
9. Cologne–Krefeld line (entire length)

Rail operations

The service is operated with push–pull trains, each composed of four double-deck carriages hauled by class 112 locomotives.

Notes

[1] "Qualitätsbericht SPNV Im Verkehrsverbund Rhein-Ruhr für 2006" (http://www.vrr.de/imperia/md/content/pressemitteilungen/ qualitaetsbericht_spnv_2006.pdf) (in German). Verkehrsverbund Rhein-Ruhr A. ö. R.. March 2007. . Retrieved 6 September 2011.

See also

- List of regional rail lines in North Rhine-Westphalia
- List of scheduled railway routes in Germany

External links

- "Rhein-Münsterland-Express" (http://nrwbahnarchiv.bplaced.net/linien/RE7.htm) (in German). *NRW rail archive*. André Joost. Retrieved 6 September 2011.

Maas-Wupper-Express

The **Maas-Wupper-Express** (RE 13) is a Regional-Express service in the German state of North Rhine-Westphalia (NRW), running from the Dutch border town of Venlo to Hamm in Westphalia.

Route

Together with the Wupper-Express (RE 4) and Rhine-Ruhr S-Bahn line S 8, the Maas-Wupper-Express provides an east-west link between the lower Rhine of Germany and the eastern Ruhr.

It runs on the tracks of the Venlo–Viersen, Viersen-Mönchengladbach, Mönchengladbach–Düsseldorf, Düsseldorf–Wuppertal, Wuppertal–Hagen and Hagen–Hamm lines.

Trains running between Venlo and Hamm have to reverse in Mönchengladbach Hauptbahnhof, so the Maas-Wupper-Express is scheduled to spend nine minutes there on the way to Venlo and ten minutes towards Hamm.

Operations

Operator of the line is Eurobahn, a subsidiary of Keolis. Operations on this line and the Rhein-Emscher-Express are carried out using 4 four-carriage and 14 five-carriage Stadler FLIRT electrical multiple units with a top speed of 160 km/h rented from Angel Trains.[1] Services run every hour.

RE 13 in Venlo

New FLIRT train

RE 13 in Mönchengladbach

Notes

[1] "Angel Trains: 18 Flirt für Keolis" (http://www.eurailpress.de/search/article/view//angel_trains_18_flirt_fuer_keolis.html) (in German). Eurailpress. 16 November 2007. . Retrieved 8 September 2011.

See also

- List of regional rail lines in North Rhine-Westphalia
- List of scheduled railway routes in Germany

External links

- "Maas-Rhein-Lippe-Netz" (http://www.maas-rhein-lippe.de/) (in German). Eurobahn. Retrieved 8 September 2011.

- "Maas-Wupper-Express" (http://nrwbahnarchiv.bplaced.net/linien/RE13.htm) (in German). *NRW rail archive*. André Joost. Retrieved 8 September 2011.

Hagen_Central_Station

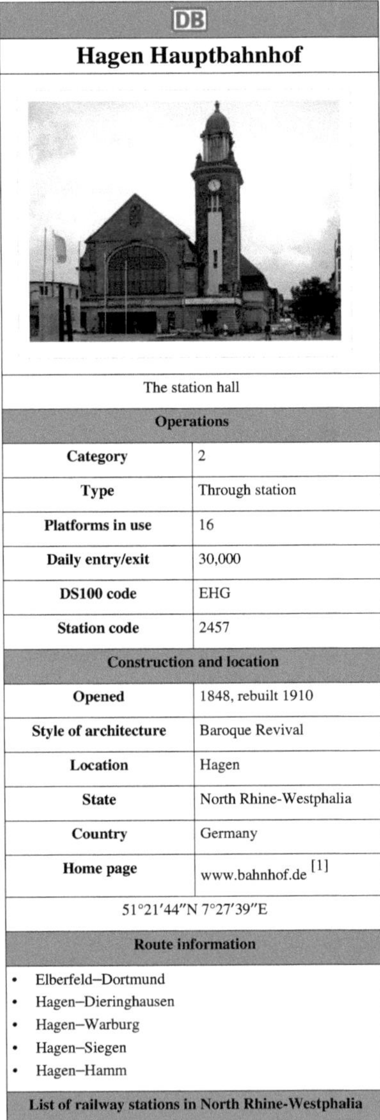

DB

Hagen Hauptbahnhof

The station hall

Operations	
Category	2
Type	Through station
Platforms in use	16
Daily entry/exit	30,000
DS100 code	EHG
Station code	2457
Construction and location	
Opened	1848, rebuilt 1910
Style of architecture	Baroque Revival
Location	Hagen
State	North Rhine-Westphalia
Country	Germany
Home page	www.bahnhof.de [1]

51°21′44″N 7°27′39″E

Route information

- Elberfeld–Dortmund
- Hagen–Dieringhausen
- Hagen–Warburg
- Hagen–Siegen
- Hagen–Hamm

List of railway stations in North Rhine-Westphalia

Hagen Central Station (German: *Hagen Hauptbahnhof*) is a railway for the city of Hagen in western Germany. It is an important rail hub for the southeastern Ruhr area, offering regional and long distance connections.

The station was opened in 1848 as part of the Bergisch-Märkische Railway Company's Elberfeld–Dortmund line and is one of the few stations in the Ruhr valley to retain its original station hall, which dates back to 1910.

History

Hagen Hauptbahnhof from the air

Thorn-Prikker: *Der Künstler als Lehrer für Handel und Gewerbe* (1911)

The original Elberfeld–Dortmund trunk line of the Bergisch-Märkische Railway Company was competed in1848/49 linking Hagen to the rapidly expanding Prussian railway network. This led to Hagen quickly becoming an industrial city based steel and metal production. After the opening of the Ruhr–Sieg railway to Siegen via Altena in 1861 the city also became an important railway junction.[2]

The Baroque Revival entrance building, opened on 14 September 1910, was built of brick and partly covered with sandstone. It survived bombing during the Second World War, although not completely, in contrast to other stations in the Ruhr area, so it can be admired today. A stained-glass window called *The Artist as Teacher of Trade and Industry* (German: *Der Künstler als Lehrer für Handel und Gewerbe*) by Jan Thorn Prikker was installed above the entrance by Karl Ernst Osthaus in 1911.[2]

Also preserved is a two-span train shed designed by Stephany from 1910. It was restored in the 1990s and is heritage-listed as an important example of a steel-constructed hall developed in the late 19th century. It is the only remaining station with a "traditional" platform area in Westphalia and the Ruhr region and one of a few of its kind in Germany. The heavy Anglo-American bombing raids in World War II on Hagen did not destroy it, unlike many other railway stations in the Ruhr.[2]

The station has points and overtaking tracks connecting to the two main platforms in the train shed. This allows up to four (short) trains to operate from each of these two-edged platforms. This has the disadvantage that passengers may sometimes be required to walk long distances.

The interior of the station was painstakingly restored from the autumn of 2004 to May 2006. Thus, the barrel vault over the concourse has been reconstructed, restoring some of its old lustre and details, including Thorn Prikker's stained-glass window, are now illuminated by daylight and are again clearly visible. This work was carried out for the 2006 World Cup of football at a total cost of €1.2 million.

The Hagen Hauptbahnhof is a listed building and is part of the The Industrial Heritage Trail (*Route Industriekultur*).[2]

Rail services

The station serves as an important link between long distance and commuter services; the InterCityExpress lines linking Cologne and Berlin call at the station as well as various InterCity and EuroCity services.

Hagen Hbf lies within the area of the Verkehrsverbund Rhein-Ruhr transport association and is served by several RegionalExpress and RegionalBahn lines as well as by two S-Bahn services of the Rhein-Ruhr S-Bahn network.

Long-distance

In long-distance passenger traffic the following Intercity and Intercity-Express services run through Hagen Hauptbahnhof:

Preceding station	Deutsche Bahn	Following station
Wuppertal *toward Aachen Hbf/Trier*	ICE 10	Hamm *toward Berlin Ostbahnhof*
Wuppertal *toward München*	ICE 31	Dortmund *toward Kiel*
Wuppertal *toward Basel SBB*	ICE 43	Dortmund *toward Hannover*
Wuppertal *toward Wien Westbf*	ICE 91	Dortmund *toward Wien Westbf*
Wuppertal *toward Passau*	IC/EC 31	Dortmund *toward Hamburg-Altona/Kiel Hbf/Puttgarden*
Wuppertal *toward Köln*	IC 55	Dortmund *toward Leipzig*

Regional trains

Preceding station	Abellio Rail	Following station
Wetter *toward Essen*	RE 16 *Ruhr-Sieg-Express*	Hagen-Hohenlimburg *toward Siegen*
Hagen-Vorhalle *toward Essen*	RB 40 *Ruhr-Lenne-Bahn*	Terminus
Terminus	RB 91 *Ruhr-Sieg - Bahn*	Hagen-Hohenlimburg *toward Siegen*
Preceding station	**Deutsche Bahn**	**Following station**
Ennepetal *toward Aachen*	RE 4 *Wupper-Express*	Witten *toward Dortmund*
Ennepetal *toward Krefeld*	RE 7 *Rhein-Münsterland-Express*	Schwerte *toward Rheine*
Terminus	RE 17 *Sauerland-Express*	Schwerte *toward Kassel*
Herdecke *toward Dortmund*	RB 52 *Volmetalbahn*	Hagen-Oberhagen *toward Lüdenscheid*
Preceding station	**eurobahn**	**Following station**

Ennepetal *toward Venlo*	RE 13 *Maas-Wupper-Express*	Schwerte *toward Hamm*
Preceding station	**Rhine-Ruhr S-Bahn**	**Following station**
Hagen-Vorhalle *toward Dortmund*	S5	*Terminus*
Hagen-Wehringhausen *toward Mönchengladbach*	S8	*Terminus*

Notes

[1] http://www.bahnhof.de/site/bahnhoefe/en/bahnhofssuche__deutschland/bahnhofssuche/
bahnhofsdaten__filter__en,variant=details,recordId=2457.html

[2] "Hagen Hauptbahnhof" (http://www.route-industriekultur.de/themenrouten/tr09/hauptbahnhof-hagen.html) (in German).
route-industriekultur. . Retrieved 7 September 2011.

External links

- "Track plan for Hagen Hauptbahnhof" (http://www.deutschebahn.com/site/bahn/de/geschaefte/
infrastruktur__schiene/netz/netzzugang/dokumente/Bahnhof/SNB/E/EHG__NBS.pdf) (in German) (PDF).
Deutsche Bahn. Retrieved 7 September 2011.
- "Hagen Hbf" (http://nrwbahnarchiv.bplaced.net/bf/8000142.htm) (in German). *NRW Rail Archive*. André
Joost. Retrieved 7 September 2011.

Remscheid_Central_Station

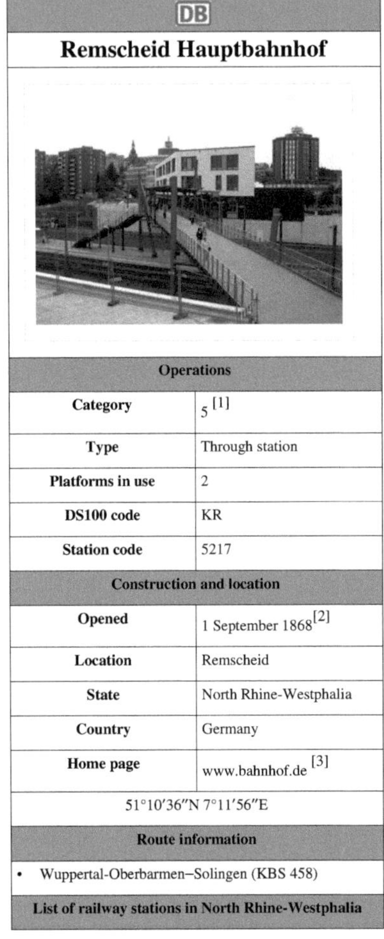

Operations	
Category	5 [1]
Type	Through station
Platforms in use	2
DS100 code	KR
Station code	5217
Construction and location	
Opened	1 September 1868 [2]
Location	Remscheid
State	North Rhine-Westphalia
Country	Germany
Home page	www.bahnhof.de [3]
51°10′36″N 7°11′56″E	
Route information	
• Wuppertal-Oberbarmen–Solingen (KBS 458)	
List of railway stations in North Rhine-Westphalia	

Remscheid Central Station (German: *Remscheid Hauptbahnhof*) is a railway station in the Bergisch city of Remscheid in the German state of North Rhine-Westphalia. It is located in Willy-Brandt-Platz near central Remscheid and is classified by Deutsche Bahn as a category 5 station.[1] . It is one of only a few Hauptbahnhöfs (central stations) in major German cities that are not served by long-distance services.

History

On 1 September 1868, the Bergisch-Märkische Railway Company (German: *Bergisch-Märkische Eisenbahn-Gesellschaft*) opened the first section of the Rittershausen–Opladen line from Oberbarmen (then called Rittershausen) to Lennep,[4] together with a branch line to Remscheid,[5] giving the city its first rail connection. Remscheid station (now called *Remscheid Hauptbahnhof*) and its attached buildings were built of timber. A branch line to Hasten was opened on 1 September 1883, and the Lennep–Remscheid line was duplicated up to 1891.[5] Five years later, a branch to Bliedinghausen was added, but has only ever been used for freight.[6]

As part of the construction of the line to Bliedinghausen the platforms were modified. Passenger trains to Hasten previously ran from their own terminal platform west of the station building. The sidings located east of Bismarckstraße later became the Remscheid East yard, which was connected to the station by a four track bridge over Bismarckstraße in 1900.[7]

During the construction of the connecting line to Solingen via Güldenwerth over the Müngsten Bridge the locomotive depot was closed due to lack of space in early 1896 and locomotive maintenance was transferred to the depot in nearby Lennep.[8]

In 1897 the gap between Solingen Süd and Remscheid was finally closed and in 1907 the route was duplicated. Thus Remscheid station now had rail connections in four directions (and also via Lennep towards Opladen). On 10 August 1911, a formal station building made of stone was inaugurated, replacing the existing building, which was called the *Zigarrenkiste* ("cigar box") because of its limited space. The line to Hasten ended again at a bay platform to the west of the station building. The main two tracks ran between a central platform and a "home" platform (next to the station building), connected by a pedestrian bridge and a tunnel for the railway post office. The new building included a rebuilt section of the hall of Elberfeld-Döppersberg station (which was closed in 1848 when Wuppertal Hauptbahnhof opened). Freight operations were moved in 1911 to the southern side of the station's track field.[7]

Until 1911, the station forecourt was called *An der Quatsche*; it was called *Adolf Müller-Strasse* during the Third Reich.[9]

In 1914,[10] Remscheid station was renamed *Remscheid Hauptbahnhof* to emphasise its importance as the most important station in the city of Remscheid. In 1922 passenger services to Hasten were discontinued.[5] Freight traffic continued with general freight traffic to Hasten and Vieringhausen freight yards as well as wagon-load traffic to ten sidings of industrial enterprises. Freight operations closed to Hasten in 1986 and to Vieringhausen in 1988; the line was officially closed on 31 December 1990. The line was dismantled between March 1993 and September 1996.[5] In 2006 the line was reconstructed as the *Trasse des Werkzeugs* ("tool path").[11]

The importance of Remscheid station was always less than the nearby Lennep station, which became the most important station when Lennep was incorporated into Remscheid in 1929.

On 28 April 1974, centralised interlocking was opened at the southwestern end of the railway land next to the Papenberger Straße level crossing.[7]

Refurbishment

After the destruction of the war, the station was rebuilt in a simplified a form on the remaining foundation and on 14 August 1956, the entrance building was re-inaugurated.[7] Only the most necessary work was carried out on the railway, the condition of which continued to deteriorate. As part of *Regionale 2006* (a regional development program of Remscheid, Solingen and Wuppertal), it was decided to carry out the complete reconstruction of the station. The old station building was demolished between December 2006 and February 2007. The old pedestrian overpass was replaced by a temporary bridge built by Technisches Hilfswerk in November 2006, which was in place until the completion of the new works. In April 2009, the new station was completed.

Old station

Rail services

Since the cessation of passenger services in 1922 to Hasten the only passenger service of the line is the service from Solingen Hauptbahnhof to Wuppertal Hauptbahnhof, now known as the Regionalbahn service *Der Müngstener* (RB 47), stopping at Remscheid.[12] Since the closure of the remaining lines branching off at Lennep, there are also no other passenger services in Remscheid city, The route from Remscheid Hauptbahnhof to Bliedinghausen is only used for freight traffic.

Two platform tracks are available for passengers and there are also several freight tracks.[13] Located on the forecourt of the station is Remscheid bus station, which is served by many urban and regional bus routes operated by Verkehrsbetriebe Remscheid and other transport companies.[12]

Preceding station	Deutsche Bahn	Following station
Remscheid-Güldenwerth *toward Solingen*	RB 47 *Der Müngstener*	Remscheid-Lennep *toward Wuppertal*

References

[1] "Station categories 2012" (http://www.deutschebahn.com/site/shared/de/dateianhaenge/infomaterial/sonstige/ bahnhofskategorieliste__db__station__service__2012.pdf) (in German) (PDF, 233.9 kb). Deutsche Bahn. . Retrieved 29 April 2012.

[2] "Remscheid Hbf operations" (http://nrwbahnarchiv.bplaced.net/kln/KR.htm) (in German). *NRW Rail Archive*. André Joost. . Retrieved 27 October 2011.

[3] http://www.bahnhof.de/site/bahnhoefe/en/bahnhofssuche__deutschland/bahnhofssuche/ bahnhofsdaten__filter__en,variant=details,recordId=5217.html

[4] "Line 2700: Wuppertal-Oberbarmen ↔ Remscheid-Lennep (↔ Opladen)" (http://nrwbahnarchiv.bplaced.net/strecken/2700.htm) (in German). *NRW Rail Archive*. André Joost. . Retrieved 28 October 2011.

[5] "Line 2705: Remscheid-Lennep ↔ Remscheid Hbf (↔ Remscheid-Hasten)" (http://nrwbahnarchiv.bplaced.net/strecken/2705.htm) (in German). *NRW Rail Archive*. André Joost. . Retrieved 28 October 2011.

[6] "Line 2706: Remscheid Hbf ↔ Remscheid-Bliedinghausen" (http://nrwbahnarchiv.bplaced.net/strecken/2706.htm) (in German). *NRW Rail Archive*. André Joost. . Retrieved 28 October 2011.

[7] Zeno Pillmann, Armin Schürings (2009). "Von Remscheid Hbf nach Hasten, Die Nebenbahn zur Filiale" (in German). *Rheinisch-Bergische Eisenbahngeschichte* (Leichlingen: A. Kaiß) (7): 21–26.

[8] Rudolf Inkeller (2009) (in German). *Das Bahnbetriebswerk Remscheid-Lennep*. **2: Bahnbetriebswerke der BD Wuppertal**. Andrea Keller Verlag, Wuppertal. p. 13. ISBN 978-3-9809930-1-2.

[9] (in German) *Remscheid − Ein verlorenes Stadtbild*.

[10] Pillmann, p. 26, others say 1 May 1925.

[11] Rudolf Inkeller (2009) (in German). *Das Bahnbetriebswerk Remscheid-Lennep*. **2: Bahnbetriebswerke der BD Wuppertal**. Andrea Keller Verlag, Wuppertal. pp. 4, 90, 98, 99. ISBN 978-3-9809930-1-2.

[12] "Remscheid Hbf" (http://nrwbahnarchiv.bplaced.net/bf/8005033.htm) (in German). *NRW Rail Archive*. André Joost. . Retrieved 27 October 2011.

[13] "Station track plan" (http://www.deutschebahn.com/site/bahn/de/geschaefte/infrastruktur__schiene/netz/netzzugang/dokumente/Bahnhof/SNB/K/KR__NBS.pdf) (in German) (PDF). Deutsche Bahn. . Retrieved 27 October 2011.

External links

- "Historical photographs of the station" (http://www.rp-online.de/bergisches-land/remscheid/nachrichten/so-sah-der-hauptbahnhof-frueher-aus-1.359407) (in German). RP Online. Retrieved 27 October 2011.

Mönchengladbach_Central_Station

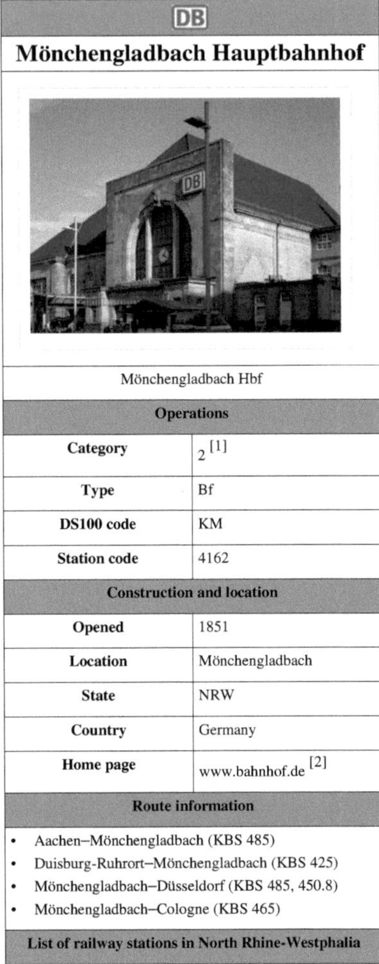

Mönchengladbach Hbf	
Operations	
Category	2 [1]
Type	Bf
DS100 code	KM
Station code	4162
Construction and location	
Opened	1851
Location	Mönchengladbach
State	NRW
Country	Germany
Home page	www.bahnhof.de [2]
Route information	
• Aachen–Mönchengladbach (KBS 485) • Duisburg-Ruhrort–Mönchengladbach (KBS 425) • Mönchengladbach–Düsseldorf (KBS 485, 450.8) • Mönchengladbach–Cologne (KBS 465)	
List of railway stations in North Rhine-Westphalia	

Mönchengladbach Central Station (German: *Mönchengladbach Hauptbahnhof*) is a railway station in the city of Mönchengladbach in western Germany.

Overview

The station is the largest railway station in the city and, along with Rheydt Hbf, one of the two Hauptbahnhof stations in Mönchengladbach. Mönchengladbach is the only city with two stations designated as a Hauptbahnhof on its soil, due to the merger between the cities of Mönchengladbach and Rheydt in the 1970s, and the subsequent reluctancy of the Deutsche Bundesbahn to rename Rheydt Hauptbahnhof. Mönchengladbach Hbf also is the largest (in terms of passengers) station in Germany to lack long-distance trains.

Railway lines calling at the station

The station is on the following routes:

- Aachen–Mönchengladbach (KBS 485)
- Duisburg-Ruhrort–Mönchengladbach (KBS 425)
- Mönchengladbach–Düsseldorf (KBS 485, 450.8)
- Mönchengladbach–Cologne (KBS 465)

Line	Line name	Route	Frequency
	Wupper-Express	Aachen Hbf–**Mönchengladbach Hbf**–Neuss Hbf–Düsseldorf Hbf–Wuppertal Hbf–Hagen Hbf–Dortmund Hbf	hourly
	Rhein-Erft-Express	**Mönchengladbach Hbf**–Köln Hbf–Köln/Bonn Flughafen–Koblenz Hbf	hourly
	Rhein-Hellweg-Express	Hamm (Westf) – Dortmund – Bochum – Essen – Duisburg – Krefeld – Viersen – **Mönchengladbach**	hourly
	Maas-Wupper-Express	Venlo–Viersen –**Mönchengladbach Hbf**–Neuss Hbf–Düsseldorf Hbf–Wuppertal Hbf–Hagen Hbf–Hamm	hourly
	Rhein-Erft-Bahn	**Mönchengladbach Hbf**–Cologne Hbf–Koblenz Hbf	Only peak hours, hourly
	Rhein-Niers-Bahn	(Wesel – Oberhausen –) Duisburg – Krefeld – Viersen – **Mönchengladbach** (– Aachen)	to Duisburg: half hourly, to Aachen and Wesel: hourly
	Schwalm-Nette-Bahn	**Mönchengladbach Hbf**–Rheydt Hbf–Wegberg–Dalheim	two hourly (peak hours: hourly)
	S 8	**Mönchengladbach Hbf**–Neuss Hbf–Düsseldorf Hbf–Wuppertal Hbf–Hagen Hbf Every 60 minutes continuing from Hagen to Dortmund as	to Wuppertal-Oberbarmen: every 20 minutes, to Hagen: every 40 minutes

Preceding station	Deutsche Bahn	Following station
Rheydt *toward Aachen*	RE 4 *Wupper-Express*	Neuss *toward Dortmund*
Rheydt *toward Koblenz*	RE 8 *Rhein-Erft-Express*	*Terminus*
Terminus	RE 11 *Rhein-Hellweg-Express*	Viersen *toward Hamm*
Rheydt *toward Koblenz*	RB 27 *Rhein-Erft-Bahn*	*Terminus*
Rheydt *toward Aachen*	RB 33 *Rhein-Niers-Bahn*	Viersen *toward Wesel*
Rheydt *toward Dalheim*	RB 39 *Schwalm-Nette-Bahn*	*Terminus*

Preceding station	eurobahn	Following station
Viersen *toward Venlo*	RE 13 *Maas-Wupper-Express*	Neuss *toward Hamm*
Preceding station	**Rhine-Ruhr S-Bahn**	**Following station**
Terminus	S8	Mönchengladbach-Lürrip *toward Hagen*

Notes

[1] "Station categories 2012" (http://www.deutschebahn.com/site/shared/de/dateianhaenge/infomaterial/sonstige/ bahnhofskategorieliste_db_station_service_2012.pdf) (in German) (PDF, 233.9 kb). Deutsche Bahn. . Retrieved 29 April 2012.

[2] http://www.bahnhof.de/site/bahnhoefe/en/bahnhofssuche_deutschland/bahnhofssuche/ bahnhofsdaten_filter_en,variant=details,recordId=4162.html

Deutsche_Bahn

Type	Aktiengesellschaft
Industry	Transport
Headquarters	Berlin, Germany
Key people	Rüdiger Grube
Products	Rail transport, Cargo transport, Services
Revenue	▲ € 37.9 billion (2011)[1]
Operating income	▲ € 2.3 billion (2011)
Net income	▲ € 1.3 billion (2011)[1]
Owner(s)	Federal Republic of Germany (100%)
Employees	229,000 (2006)[1]
Subsidiaries	Arriva BAX Global Chiltern Railways DB Fernverkehr DB Regio DB Station&Service Railion Schenker AG CrossCountry
Website	www.deutschebahn.com [2]

Deutsche Bahn AG (*DB AG*, *DBAG* or *DB*) is the German national railway company, a private joint-stock company (AG) with its headquarters in Berlin.[3] It came into existence in 1994 as the successor to the former state railways of Germany, the Deutsche Bundesbahn of West Germany and the Deutsche Reichsbahn of East Germany. It also gained ownership of former railway assets in West Berlin held by the VdeR. Its name means "German Railway" in German.

DB is organized as a business group and has over 1000 subsidiaries, of which 287 are in Germany.[4] It describes itself as the second-largest transport company in the world after Deutsche Post AG and is the largest railway operator and infrastructure owner in Europe. It carries about two billion passengers each year.

DBAG has taken over the abbreviation and logo *DB* from the West German state railway Deutsche Bundesbahn, although it has modernised the logo. Erik Spiekermann designed the new corporate font DB type.

Originally, DBAG had its headquarters in Frankfurt am Main but moved to Potsdamer Platz in central Berlin in 1996, where it occupies a 26-story office tower designed by Helmut Jahn at the eastern end of the Sony Center and named BahnTower. As the lease was to expire in 2010, DB had announced plans to relocate to Berlin Hauptbahnhof, and in 2007 a proposal for a new headquarters by 3XN Architects won an architectural competition which also included Foster + Partners, Dominique Perrault and Auer + Weber.[5] However, these plans have been put on hold, and the BahnTower leased for at least three more years.[6]

History

While the railway network in Germany dates back to 1835 when the first tracks were laid on a 6 km (**unknown operator: u'strong'** mi) route between Nuremberg and Fürth, Deutsche Bahn has been a relatively recent development in German railway history. Founded in January 1994 as a joint-stock company, Deutsche Bahn was designed to operate the railways of both the former East and West Germany after unification in November 1989 as a single, uniform, privately-run company.[7] There are three main periods of development in this unified German railway: its formation, its early years (1994–1999), and the period from 1999 to the present.

1999 to present

The second step of the *Bahnreform* (Railway reform) was carried out in 1999. All rolling stock, track, personnel and real assets were divided between the holding company and the five principal subsidiaries of DBAG: *DB Reise & Touristik AG* (long distance passenger service, later renamed *DB Fernverkehr AG*), *DB Regio AG* (regional passenger services, in the course of the reform under charge of the federal states), *DB Cargo AG* (freight services, later changed to *Railion AG*), *DB Netz AG* (operating the railway system), and *DB Station & Service AG* (operating the stations). This new organisational scheme was introduced not least to implement European Community directive 91/440/EEC that demands access to railway systems free of discrimination.

In December 2007, DB reorganised again, bringing all passenger services into its DB Bahn arm, logistics under DB Schenker and infrastructure and operations under DB Netze.

Corporate subdivisions

The DB group (Deutsche Bahn AG) is divided into four main operations groups: Arriva, DB Bahn, DB Netze and DB Schenker. These subsidiaries are companies in their own right, although most of them are 100% owned by DBAG.[1]

DB building, across the street from the Sony Center in Potsdamer Platz in Berlin.

Arriva

Deutsche Bahn placed a bid in May 2010 for the UK-based transport company Arriva. Arriva runs bus and rail companies in 12 European countries. The merger was approved by the European Commission in August 2010, subject to DB divesting Arriva services in Germany (these are now run as Netinera). The merger became effective on 27 August 2010.[8]

Services in the UK formerly run as DB Regio are now operated by a new subdivision of the company, Arriva UK Trains.

Including Arriva Trains Wales, Cross Country Trains and Chiltern Trains.

Logo Mark

DB Bahn

DB Bahn is the group that manages passenger travel within Germany.

Originally called *Reise & Touristik*, this group is responsible for the managing, ticketing, servicing and running of all German *Intercity-Express*, *EuroCity*, *Intercity* and *Regionalbahn* trains, and many commuter-oriented urban *Stadtschnellbahn* (generally abbreviated as S-Bahn) networks within Germany. The group also handles the information and customer service side of the operation.[1]

This group is divided into three business areas: DB Fernverkehr, DB Regio and DB Stadtverkehr.

DB Fernverkehr

DB Fernverkehr AG is a semi-independent division of Deutsche Bahn that operates long-distance passenger trains in Germany. It was founded in 1999 in the second stage of the privatisation of German Federal Railways under the name of DB Reise&Touristik and renamed in 2003.

DB Fernverkehr operates all InterCityExpress and InterCity trains in Germany as well as several EuroCity trains throughout Europe. Unlike its sister companies DB Regio and Railion (formerly DB Cargo), DB Fernverkehr still holds a de-facto monopoly in its segment of the market as it operates hundreds of trains per day, while all competitors' long-distance services (Veolia Verkehr most notably) combined amount to no more than 10-15 trains per day.

Deutsche Bahn ICE 3 high speed train on the Frankfurt-Cologne high-speed rail line, Germany

DB Regio

DB Regio AG is the subsidiary of Deutsche Bahn that operates passenger trains on short and medium distances in Germany. Unlike its long-distance counterpart, DB Fernverkehr, it does not operate trains on its own account. Traffic is ordered and paid for by the Bundesländer or their Landkreise. Competition for those state-sponsored services is somewhat more fierce than for long-distance services. Consequently, DB Regio has lost a considerable number of routes to its competitors. Some states have awarded long-term contracts to DB Regio (usually 10 to 15 years), which has sparked harsh criticism from anti-trust groups.

DB Stadtverkehr

DB Stadtverkehr was responsible for commuter services of the Berlin and Hamburg S-Bahns and numerous bus companies (see Bahnbus). The subsidiary was integrated into DB Regio on 31 December 2010. The two S-Bahn networks handle over 500 million passengers annually.[9]

DB Netze

Since the end of 2007 DB Netze has been responsible for infrastructure and operations, taking over from DB Netz AG. Its business areas including DB Netze Fahrweg, DB Netze Energie, DB Netze Personenbahnhöfe, DB ProjektBau and DB Station&Service. A further business area known as DB Dienstleistungen covers six different areas of operations: DB Fahrzeuginstandhaltung, DB Systel, DB Services, DB Fuhrpark, DB Kommunikationstechnik and DB Sicherheit.

DB Schenker

DB Schenker is the logistics arm of DB, as of 2008 it employed over 88000 people, and was the largest European rail freight company.[10]

Its two business areas are DB Schenker Rail (formerly Railion.[11]) and DB Schenker Logistics. Other subsidiaries include Bax Global, Transfesa and the former EWS, now DB Schenker Rail (UK) Ltd.. DB Schenker Rail has its head office in Mainz, and is the holding company for the five (at 1 January 2009) national subsidiaries Railion Deutschland, Railion Nederland, Railion Danmark, Railion Italy and Railion Schweiz.

Foreign firms

DB also has interests abroad, owning rail freight company EWS in the UK, which also operates the British Royal Train[11] and also has interests in Eastern Europe.

It is possible to obtain train times for any journey in Europe from Deutsche Bahn's website.[12]

Trans-Eurasia Logistics is a joint venture with Russian Railways (RZD) that operates container freight trains between Germany and China via Russia.

Members of the board

- Dr. Richard Lutz (Finance & Controlling)[1]
- Ulrich Weber (Human resources)[1]
- Dr. Volker Kefer (Technics, Railway interconnection and Services)[1]
- Dr. Karl-Friedrich Rausch (Transportations and LogisticsPassenger traffic)
- Ulrich Homburg (Passenger transport)[1]
- Dr. Volker Kefer (Infrastructure and services)[1]

Dr Werner Müller is the current director of the supervisory board (also at Degussa and Ruhrkohle AG).[1]

Previous chairmen of the board were

- Heinz Dürr, 1994–1997, became director of the supervisory board afterwards.[1]
- Johannes Ludewig, 9 July 1997-September 30, 1999[1]

Codeshare with airlines

In conjunction with American Airlines, Emirates Airline, China Airlines, TAM Airlines and Lufthansa, *Deutsche Bahn* operates the AiRail Service between Frankfurt International Airport and Cologne / Bonn, Düsseldorf, Freiburg, Hamburg, Hanover, Mannheim, Munich, Nuremberg, and Stuttgart. DB has the IATA designator 2A.[1]

Corporate headquarters (the BahnTower) at Potsdamer Platz in Berlin, Germany

Privatization project

The Social-Democrat Minister Wolfgang Tiefensee was to present a bill before the council of ministers which envisaged selling 25% of DB, beginning in 2008. At term, the state should retain control by owning a 51% stake. Deutsche Bahn AG is evaluated at €45 billion.[13] However, the railways are evaluated at €200 billion.[14]

In October 2007, the train drivers' strike was another problem for the privatization plans. The drivers' union GDL refused to accept the labour contracts between DB and other unions, claiming a 31% pay rise for its members. The strike was the first nationwide railway strike since 1992.[15]

See also

- DB Fernverkehr
- DB Regio
- DB NachtZug
- Bahn TV
- Railway electrification system
- Transport in Germany
- Rail transport in Germany

References

[1] Deutsche Bahn AG. "Daten und Fakten zum Geschäftsbericht 2005" (http://www.db.de/site/shared/de/dateianhaenge/berichte/
daten__und__fakten__2005.pdf). . Retrieved 2006-12-17.

[2] http://www.deutschebahn.com/

[3] " Deutsche Bahn AG at a glance (http://www.deutschebahn.com/site/bahn/en/db__group/corporate__group/ata__glance/facts__figures/
facts__figures.html)." *Deustche Bahn*. Retrieved 12 May 2009.

[4] eurailpress.de (http://www.eurailpress.de/article/view/4/deutsche-bahn-raeumt-bei-tochtergesellschaften-auf.html)

[5] "Competition win for 3XN" (http://www.worldarchitecturenews.com/index.php?fuseaction=wanappln.projectview&upload_id=1766).
World Architecture News. 21 December 2007. . Retrieved 2009-10-33.

[6] http://www.ftd.de/unternehmen/handel_dienstleister/:Deutsche%20Bahn%20Pl%E4ne%20Umzug/346601.html

[7] "The foundation of Deutsche Bahn AG" company website (http://www.deutschebahn.com/site/bahn/en/db__group/corporate__group/
history/topics/foundation/foundation.html)

[8] "EC approves DB's takeover of Arriva" (http://www.railwaygazette.com/news/single-view/view/10/ec-approves-dbs-takeover-of-arriva/
browse/2.html). *Railway Gazette International* (London). 11 August 2010. . Retrieved 2010-08-19.

[9] Deutsche Bahn AG: *Kennzahlen 2005 — DB AG mit bestem Geschäftsjahr ihrer Geschichte* (http://www.db.de/site/bahn/de/
unternehmen/investor__relations/kennzahlen/kennzahlen__2005.html), 23 September 2006

[10] "Transportation and logistics in the DB schenker group - Profile" (http://web.archive.org/web/20080828025328/http://www.
dbschenker.com/site/logistics/dbschenker/com/en/about__dbschenker/profile/profile__alt.html), *www.dbschenker.com*, archived from
the original (http://www.dbschenker.com/site/logistics/dbschenker/com/en/about__dbschenker/profile/profile__alt.html) on 28 Aug
2008,

[11] Macalister, Terry (28 June 2007). "Deutsche Bahn to run Queen's train" (http://www.guardian.co.uk/business/2007/jun/28/
transportintheuk). *guardian.co.uk (web only)*. .

[12] Tickets - Timetable (http://www.bahn.de/p/view/international/englisch/international_guests.shtml)

[13] "L'Allemagne s'apprête à vendre 25 % de la Deutsche Bahn" (http://www.lemonde.fr/web/article/
0,1-0@2-3234,36-938654@51-925598,0.html) (in French). *Le Monde* (Paris). 24 July 2007. .

[14] Deutscher Bundestag Drucksache 16/3801 (http://dip21.bundestag.de/dip21/btd/16/038/1603801.pdf)

[15] "German court blocks train strike" (http://news.bbc.co.uk/2/hi/business/6936778.stm). *BBC News*. 8 August 2007. .

External links

- DB Corporate Home Page (http://www.deutschebahn.com/site/bahn/en/start.html)
- DB Corporate Home Page (http://www.deutschebahn.com) (German)
- DB travel portal (http://www.bahn.com)
- DB timetable information (http://reiseauskunft.bahn.de/bin/query.exe/en)
- Pictures of German train and Hamburg (http://www.hamburg-hautnah.de/db.html)

Wuppertal_Central_Station

Wuppertal Hauptbahnhof	
Entrance building	
Operations	
Category	2
Type	Through station
Platforms in use	5
DS100 code	KW
Station code	6914
Construction and location	
Opened	1850 [1]
Style of architecture	Neoclassical
Architect	Hauptner and Ebeling
Location	Wuppertal
State	North Rhine-Westphalia
Country	Germany
51°15′17″N 7°9′0″E	
Route information	
• Düsseldorf–Wuppertal • Wuppertal–Dortmund • (Wuppertal–)Gruiten–Köln-Deutz • (Wuppertal–)Wuppertal-Oberbarmen–Solingen • (Wuppertal–)Wuppertal-Vohwinkel–Essen	
List of railway stations in North Rhine-Westphalia	

Wuppertal Central Station (German: *Wuppertal Hauptbahnhof*) is a railway station for the city of Wuppertal, which is just south of the Ruhr Area, in the German state of North Rhine-Westphalia. It is on the line between Düsseldorf/Cologne and Dortmund. The 1848 reception building is one of the oldest of its kind. The station was originally Elberfeld station and has been renamed several times since. Since 1992, it has been called *Wuppertal Hauptbahnhof.*[1]

History

The Bergisch-Märkische station in 1855, lithography by Wilhelm Riefstahl

The smallest metropolitan station in Germany

Station forecourt

S-Bahn

On 3 September 1841, a few years after the opening of the first railway in Germany, the Dusseldorf-Elberfeld Railway Company (German: *Düsseldorf-Elberfelder Eisenbahn-Gesellschaft*, DEE) began operation of the Düsseldorf–Elberfeld line from its Düsseldorf station to its Elberfeld station (now Wuppertal-Steinbeck station).[2] It was the first steam-worked railway line in Western Germany and Prussia.

The Bergisch-Märkische Railway Company (*Bergisch-Märkische Eisenbahn-Gesellschaft*, BME), opened its Elberfeld–Dortmund railway from its Elberfeld station (known as Döppersberg station) via Hagen to Dortmund to Schwelm on 9 October 1847. It was extended to Hagen and Dortmund on 20 December 1848.[2] The BME took over the DEE in 1857.

The first provisional station building became inadequate within a few years. It was decided to build a new building, designed by Hauptner and Ebeling and opened in 1850[1] on a new section of line connecting the BME and DEE lines, which was completed on 9 March 1849.[2] Around 1900, a protruding porch was built in front of the ground floor, which conflicted with the architectural design. Nevertheless, this concept was maintained after its reconstruction after World War II. This will only change with the completion of the current renovation of the station/Döppersberg area.

The station has been renamed several times. It was first called *Elberfeld*, but a few years later it was renamed *Elberfeld-Döppersberg* and before the First World War it was renamed *Elberfeld Hauptbahnhof*. In the early 1930s the station's name was changed to *Wuppertal-Elberfeld* station as a consequence of the merger of the towns of Elberfeld and Barmen as the city of Wuppertal. Finally in 1992, it was renamed *Wuppertal Hauptbahnhof*.[1]

Station Buildings

The station building is located next to platform track 1 and is connected by a tunnel to tracks 2–5. Above the entrance, near the old Reichsbahn railway division of Elberfeld, there are four pillars supporting the roof. The building is connected by the 200 metre long Döppersberg pedestrian tunnel directly with central Elberfeld and the *Wuppertal Hbf (Döppersberg)* Schwebebahn (monorail) station.

A McDonald's restaurant has been established in the premises of the former baggage check-in and in the tunnel under the entrance there a large newsagency/book shop and a bakery. The low building in front of the historic station building houses a pharmacy. In front of the entrance to the station there is a parking area, including a taxi stand, and nearby there is a Inter City Hotel.

Architecture

The original building is one of the oldest big city railway stations in Germany. It is a three storey ashlar building bounded by tower-like corner projections. The main entrance in the middle of the building is a four-columned portico, with emphasised Corinthian capitals and has strong antique ornamentation. The ground floor originally had arched openings and it has six rectangular windows on each level and on each side of the portico. It was necessary in 1900 to build a ground-floor entrance porch to cater for the growing need for space for counters and waiting rooms.

The station is part of a ensemble of buildings built in neoclassical style, which is grouped around the railway station forecourt. On the western side of the square is the headquarters of the former Reichsbahn railway division of Elberfeld; on the eastern side there used to be the headquarters of the Chief General Manager, but thus was torn down after the Second World War.

The construction of the station was accompanied by extensive urban development in the Döppersberg area. The Döppersberg bridge (*Döppersberger Brücke*) was built to connect centre of old Elberfeld with the station over the Wupper.

Reconstruction

Preparations for the reconstruction of the Hauptbahnhof and the surrounding area of Döppersberg began in July 2009. The modernisation of the station was formally launched on 30 June 2009. The new station will have a two-floor shopping level, the "Mall", a square glass cube with space for offices, a large station forecourt, built in the current Bahnhofstraße, and a bridge, which will include a café, over federal highway 7 (B 7), which will be lowered by about seven metres. The bus station that is now on the B 7 will replace the car park next to the station. The renovation will be completed in 2016.

The modernisation of the entrance building by Deutsche Bahn will start in 2014, at a total cost of € 12.4 million. It is expected that these upgrades will be completed in 2016 simultaneously with the reconstruction of the Döppersberg area.[3]

Train services

Although the station possesses only five tracks, less than the other stations of the city, nearly all services running through Wuppertal stop here, except for the S 68 S-Bahn service terminating in Vohwinkel. The following services stop at the station:

- S-Bahn (S 8, S 9)
- Regionalbahn (RB47, RB48)
- Regionalexpress (RE4, RE7, RE13)
- InterCity
- InterCityExpress
- EuroCity

Long distance trains

The following services currently call at Wuppertal Hauptbahnhof:

Series	Operator	Route	Material	Frequency	Notes
ICE31	DB	Dortmund Hbf - Hagen Hbf - Wuppertal Hbf - Solingen Hbf - Köln Hbf - Siegburg/Bonn - Frankfurt/Main Flughafen Fernbahnhof - Mannheim Hbf - Karlsruhe Hbf - Offenburg - Freiburg Hbf - Basel Bad Bf - Basel SBB	3x per day		
IC55	DB	Köln Hbf - Solingen Hbf - Wuppertal Hbf - Hagen Hbf - Dortmund Hbf - Hamm (Westf) - Gütersloh Hbf - Bielefeld Hbf - Herford - Bad Oeynhausen - Minden (Westf) - Hannover Hbf - Wolfsburg Hbf - Magdeburg Hbf - Köthen - Halle (Saale) Hbf - Leipzig/Halle Flughafen - Leipzig Hbf		Every 2 Hours	

Preceding station	Deutsche Bahn	Following station
Köln *toward München*	ICE 31	Hagen *toward Kiel*
Solingen *toward Basel SBB*	ICE 43	Hagen *toward Hannover*
Solingen *toward Wien Westbf*	ICE 91	Hagen *toward Wien Westbf*
Solingen *toward Passau*	IC/EC 31	Hagen *toward Hamburg-Altona/Kiel Hbf/Puttgarden*
Solingen *toward Köln*	IC 55	Hagen *toward Leipzig*

Regional trains

The following regional services (Regional-Express and Regionalbahn) call at Wuppertal Hbf:[4]

- *Wupper-Express* Dortmund - Hagen - Wuppertal - Düsseldorf - Mönchengladbach - Aachen
- *Rhein-Münsterland-Express* Rheine - Münster - Hagen - Wuppertal - Cologne - Krefeld
- *Maas-Wupper-Express* Venlo - Viersen - Mönchengladbach - Düsseldorf - Wuppertal - Hagen - Hamm
- *Rhein-Wupper-Bahn* Bonn-Mehlem - Bonn - Cologne - Solingen - Wuppertal

Series	Operator	Route	Material	Frequency	Notes
RB47 *Der Müngstener*	DB	Wuppertal Hbf - Wuppertal-Unterbarmen - Wuppertal-Barmen - Wuppertal-Oberbarmen - Wuppertal-Ronsdorf - Remscheid-Lüttringhausen - Remscheid-Lennep - Remscheid Hbf - Remscheid-Güldenwerth - Solingen-Schaberg - Solingen Mitte - Solingen Grünewald - Solingen Hbf	DB Class 628	Every 20 Minutes	

Preceding station	Deutsche Bahn	Following station
Wuppertal-Vohwinkel *toward Aachen*	RE 4 *Wupper-Express*	Wuppertal-Barmen *toward Dortmund*
Solingen *toward Krefeld*	RE 7 *Rhein-Münsterland-Express*	Wuppertal-Oberbarmen *toward Rheine*
Terminus	RB 47 *Der Müngstener*	Wuppertal-Unterbarmen *toward Solingen*
Wuppertal-Vohwinkel *toward Bonn-Mehlem*	RB 48 *Rhein-Wupper-Bahn*	*Terminus*
Preceding station	**eurobahn**	**Following station**
Wuppertal-Vohwinkel *toward Venlo*	RE 13 *Maas-Wupper-Express*	Wuppertal-Barmen *toward Hamm*
Preceding station	**Rhine-Ruhr S-Bahn**	**Following station**
Wuppertal-Steinbeck *toward Mönchengladbach*	S8	Wuppertal-Unterbarmen *toward Hagen*
Wuppertal-Steinbeck *toward Haltern am See*	S9	*Terminus*

References

[1] "Wuppertal Hauptbahnhof operations" (http://nrwbahnarchiv.bplaced.net/kln/KW.htm) (in German). *NRW Rail Archive*. André Joost. . Retrieved 31 October 2011.

[2] "Line 2550: Aachen - Kassel" (http://nrwbahnarchiv.bplaced.net/strecken/2550.htm) (in German). *NRW Rail Archive*. André Joost. . Retrieved 31 October 2011.

[3] "Döppersberg: Ab 2014 wird der Bahnhof saniert" (http://www.google.com.au/webhp?sourceid=navclient&hl=en-GB&ie=UTF-8& rlz=1T4ASUS_enAU306AU309) (in German). *Westdeutsche Zeitung*. . Retrieved 3 November 2011.

[4] "Wuppertal Hauptbahnhof" (http://nrwbahnarchiv.bplaced.net/bf/8000266.htm) (in German). *NRW Rail Archive*. André Joost. . Retrieved 31 October 2011.

External links

- "Wuppertal Hauptbahnhof track plan" (http://www.deutschebahn.com/site/bahn/de/geschaefte/ infrastruktur__schiene/netz/netzzugang/dokumente/Bahnhof/SNB/K/KW__NBS.pdf) (in German) (PDF, 319 kB). Deutsche Bahn. Retrieved 31 October 2011.

German_railway_station_categories

About 5,400 railway stations in Germany that are owned and operated by the Deutsche Bahn subsidiary DB Station&Service are assigned into **seven categories**, denoting the service level available at the station.

Their assignment into the categories also influences the amount of money railway companies need to pay to DB Station&Service for using the facilities at the stations.

Categories

Berlin Hbf (category 1)

Category 1

The 20 stations of Category 1 are considered traffic hubs. They are permanently staffed and carry all sorts of railway-related facilities as well as usually featuring a shopping mall in the station. Most of these stations are the main stations (*Hauptbahnhof* or *Hbf*) of large cities with 500,000 inhabitants and above, though some in smaller cities, like Karlsruhe Hauptbahnhof, are regarded as important because they are at the intersection of important railway lines. Berlin, Hamburg, Munich and Cologne, the four biggest cities in Germany, have more than one Category 1 station. Included in the category are the following stations:

Hamburg Hbf (category 1)

Hamm (Westf) (category 2)

- Berlin-Gesundbrunnen station
- Berlin Hauptbahnhof
- Berlin Ostbahnhof
- Berlin Südkreuz
- Dortmund Hauptbahnhof
- Dresden Hauptbahnhof
- Düsseldorf Hauptbahnhof
- Essen Hauptbahnhof
- Frankfurt (Main) Hauptbahnhof
- Hamburg-Altona station
- Hamburg Hauptbahnhof
- Hannover Hauptbahnhof
- Karlsruhe Hauptbahnhof
- Köln Hauptbahnhof
- Köln Messe/Deutz
- Leipzig Hauptbahnhof
- München Hauptbahnhof
- München Ost station
- Nürnberg Hauptbahnhof
- Stuttgart Hauptbahnhof

Category 2

Most of the about 80 stations of Category 2 are either important junctions of long-distance traffic or offer connections to large airports. InterCity and EuroCity trains generally call at these stations. All railway-related services, like a ticket hall and a service desk, are present at the station and the station is staffed during the usual times of traffic. The service is similar to Category 1 stations. The Category 2 stations, by state, are:

Lichtenfels (category 3)

- Baden-Württemberg (8): Bietigheim-Bissingen, Freiburg (Brsg) Hbf, Heidelberg Hbf, Mannheim Hbf, Offenburg, Plochingen, Tübingen Hbf, Ulm Hbf
- Bavaria (7): Aschaffenburg Hbf, Augsburg Hbf, Bamberg, Fürth Hauptbahnhof, München-Pasing, Regensburg Hbf, Würzburg Hbf
- Berlin (6): Friedrichstraße, Lichtenberg, Potsdamer Platz, Spandau, Wannsee, Zoologischer Garten
- Brandenburg (2): Cottbus, Potsdam Hbf
- Bremen (1): Bremen Hbf
- Hamburg (2): Hamburg Dammtor, Hamburg-Harburg
- Hesse (8): Darmstadt Hbf, Frankfurt (Main) Süd, Fulda, Gießen, Hanau Hbf, Kassel-Wilhelmshöhe, Kassel Hbf, Wiesbaden Hbf

Montabaur (category 4)

- Lower Saxony (6): Braunschweig Hbf, Göttingen, Hildesheim Hbf, Oldenburg (Oldb) Hbf, Osnabrück Hbf, Wolfsburg Hbf
- Mecklenburg-Vorpommern (1): Rostock Hbf
- North Rhine-Westphalia (15): Aachen Hauptbahnhof, Altenbeken, Bielefeld Hbf, Bochum Hbf, Bonn Hbf, Duisburg Hbf, Düsseldorf Flughafen, Gelsenkirchen Hbf, Hagen Hbf, Hamm (Westf), Mönchengladbach Hbf, Münster (Westf) Hbf, Neuss Hbf, Solingen Hbf, Wuppertal Hbf
- Rhineland-Palatinate (7): Kaiserslauten Hbf, Koblenz Hbf, Ludwigshafen Hbf, Mainz Hbf, Neustadt (Weinstraße) Hbf, Trier Hbf, Worms Hbf

Köln-Holweide (category 5)

- Saarland (1): Saarbrücken Hbf
- Saxony (2): Chemnitz Hbf, Dresden-Neustadt
- Saxony-Anhalt (2): Halle (Saale) Hbf, Magdeburg Hbf
- Schleswig-Holstein (3): Kiel Hbf, Lübeck Hbf, Neumünster
- Thuringia (2): Erfurt Hbf, Weimar

Category 3

230 stations belong to Category 3. These stations will usually feature a station hall where travellers can buy tickets and groceries, but these stations are usually not permanently staffed. They are often main stations of cities with about 50,000 inhabitants.

Examples are Lichtenfels, Passau Hbf or Mülheim (Ruhr) Hbf.

Hagen-Vorhalle (category 6)

Category 4

Category 4 includes 600 stations. Most of these stations have frequent connections with RegionalExpress and RegionalBahn trains. Their service level is comparable to a bus station and they offer services to commuters. This category also includes stations situated in major cities that see high usage of S-Bahn or RE/RB services.

Duisburg-Entenfang station (category 7)

Examples are Montabaur or Coburg.

Category 5

1,070 stations make up Category 5. These stations either belong to smaller, rural towns or to outlying suburban areas of major cities. Their inventory normally is "vandal-proofed" due to the lower number of passengers. They normally only have local trains calling at the station.

Examples are Köln-Holweide or Bremerhaven-Lehe.

Category 6

Category 6 includes 2,500 stations. These stations have low passenger numbers and only the most basic equipment needed is present at the station. They are akin to bus stops.

Examples of this category are Loxstedt or Hagen-Vorhalle.

Category 7

Most of the 870 stations of the lowest category 7 are in rural areas. Only several local trains call at these stops, which usually have not more than one platform. An example for this category is Duisburg-Entenfang.

References

"Die sieben Bahnhofskategorien" [1]. DB Station&Service AG. Retrieved 2011-05-15.

External links

- "Station categories 2012" [2] (in German) (PDF, 233.9 kb). Deutsche Bahn. Retrieved 29 April 2012.

References

[1] http://www.db.de/site/bahn/de/geschaefte/infrastruktur__schiene/personenbahnhoefe/bahnhofskategorien/bahnhofs__kategorien.html
[2] http://www.deutschebahn.com/site/shared/de/dateianhaenge/infomaterial/sonstige/
      bahnhofskategorieliste__db__station__service__2012.pdf

Wuppertal_Schwebebahn

<table>
<tr><td colspan="2" align="center">Wuppertal Schwebebahn</td></tr>
<tr><td colspan="2" align="center">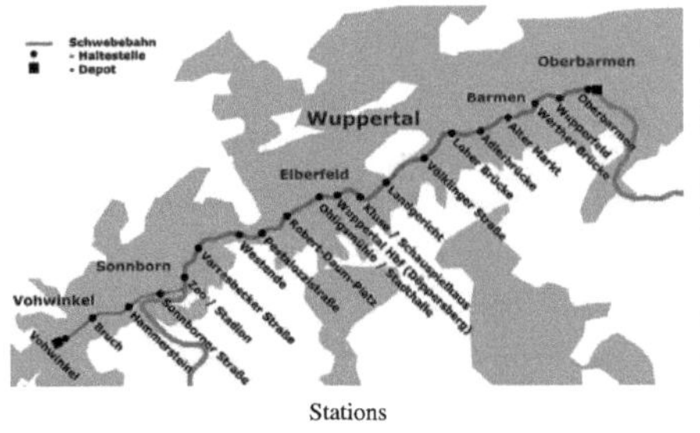
Wuppertal Schwebebahn in 2010</td></tr>
<tr><td colspan="2" align="center">Background</td></tr>
<tr><td>Locale</td><td>Wuppertal, Germany</td></tr>
<tr><td>Transit type</td><td>suspended monorail</td></tr>
<tr><td>Number of lines</td><td>1</td></tr>
<tr><td>Number of stations</td><td>20</td></tr>
<tr><td>Daily ridership</td><td>82,000[1]</td></tr>
<tr><td colspan="2" align="center">Operation</td></tr>
<tr><td>Began operation</td><td>1 March 1901</td></tr>
<tr><td>Operator(s)</td><td>Wuppertaler Stadtwerke (WSW)</td></tr>
<tr><td colspan="2" align="center">Technical</td></tr>
<tr><td>System length</td><td>13.3 kilometres (8.3 miles)</td></tr>
</table>

Wuppertal Schwebebahn or *Wuppertal Floating Tram* is a suspension railway in Wuppertal, Germany. Its full name is "Electric Highway Installation (Floating Tram), Eugen Langen System" (*Anlage einer elektrischen Hochbahn (Schwebebahn), System Eugen Langen*).[2] Designed by Eugen Langen to be used in Berlin, the installation with elevated stations was built in Barmen, Elberfeld and Vohwinkel between 1897 and 1903, the first track opened in 1901. The Schwebebahn is still in use today as normal means of local public transport, moving 25 million passengers annually (2008).[3]

Stations

The suspension railway travels along a route of 13.3 kilometres (**unknown operator: u'strong'** mi), at a height of about 12 metres () above the river Wupper between Oberbarmen and Sonnborner Straße (10 kilometres (**unknown operator: u'strong'** mi)) and about 8 metres () above the valley road between Sonnborner Straße and Vohwinkel (3.3 kilometres (**unknown operator: u'strong'** mi)).[4] [5] At one point the railway crosses the A46 motorway. The entire trip takes about 30 minutes.[5] The Schwebebahn operates within the VRR transport association and accepts tickets issued by the VRR companies.

History

The Wuppertal Schwebebahn had a forerunner: in 1824, Henry Robinson Palmer of England presented a railway system which differed from all previous constructions. It was basically a low single-rail suspension railway on which the carriages were drawn by horses. Friedrich Harkort, a Prussian industrial entrepreneur and politician, loved the idea. He saw big advantages for the transportation of coal to the early industrialised region in and around the Wupper valley. Harkort had his own steel mill in Elberfeld; he built a demonstration segment of the Palmer system and set it up in 1826 on the grounds of what is today the Wuppertal tax office. He therefore tried to attract public attention to his railway plans.

On 9 September 1826, the town councillors of Elberfeld met to discuss the use of a "Palmer's Railway" from the Ruhr region, Hinsbeck or Langenberg, to the Wupper valley, Elberfeld, connecting Harkort's factories. Friedrich Harkort inspected the projected route with a surveyor and a member of the town council. The plans never went ahead, due to protests from the transport branch and owners of mills that were not on the routes.

In 1887 the cities of Elberfeld and Barmen formed a commission for the construction of an elevated railway or *Hochbahn*. In 1894 they chose the system of the engineer Eugen Langen of Cologne, and in 1896 the order was licensed by the City of Düsseldorf.[5] [6] In 2003, the Rhine Heritage Office (*Rheinische Amt für Denkmalpflege des Landschaftsverbandes Rheinland* or LVR) announced the discovery of an original section of the test route of the Wuppertal Schwebebahn.

Construction on the actual Wuppertal Schwebebahn began in 1898, overseen by the government's master builder, Wilhelm Feldmann. On 24 October 1900, William II, German Emperor participated in a monorail trial run.[5]

In 1901 the railway came into operation. It opened in sections: the line from Kluse to Zoo/Stadion opened on 1 March, the line to the western terminus at Vohwinkel opened on 24 May, while the line to the eastern terminus at Oberbarmen did not open until 27 June 1903.[4] Around 19,200 tonnes of steel were used to produce the supporting frame and the train stations.[5] The construction cost 16 million Goldmark. Since its first opening, the railway has closed once due to severe damage from World War II, but managed to open as early as 1946.[4]

Construction of Wuppertaler Schwebebahn, 1900

Station Werther Brücke 1913

A 1977 Schwebebahn, crossing an intersection

Current modernisation

The Wuppertal Schwebebahn nowadays carries up to 82,000[1] passengers a day through the city. Since 1997, the supporting frame has been largely modernised, and many stations have been reconstructed and brought technically up to date. The "Kluse" station, at the theatre in Elberfeld, had been destroyed during the Second World War; this too was reconstructed during the modernisation. Work was planned to be completed in 2001; however a serious accident took place in 1999 which left five people dead and 47 injured. This, along with delivery problems, meant that the completion date was delayed. In recent years (2004), the cost of the reconstruction work has increased from €380 million to €480 million.[7]

Since 2004, many of the railway stations have been fitted with CCTV cameras.

On 15 December 2009 the Schwebebahn temporarily suspended its operations for safety concerns; several of the older support structures needed to be renewed, – a process that was completed on 19 April 2010.[8]

On 10 November 2011 Wuppertaller Stadtwerke signed a contract with Vossloh Kiepe to supply 31 new articulated cars to replace those built in the 1970s. These will be built by Vossloh Rail Vehicles in Valencia, Spain. When these cars are introduced the line's power supply will be uprated from 600 to 750v.[9]

Stations

- Oberbarmen – eastern terminus
- Wupperfeld
- Werther Brücke
- Alter Markt
- Adlerbrücke
- Loher Brücke
- Völklinger Street
- Landgericht
- Kluse
- Hauptbahnhof
- Ohligsmühle
- Robert-Daum-Platz
- Pestalozzistr
- Westende
- Varresbecker Street
- Zoo/Stadion
- Sonnborner Straße
- Hammerstein
- Bruch
- Vohwinkel – western terminus

Station Zoo/Stadion

Station Sonnborner Straße

Technology

The tram wagons are suspended from a single rail built underneath a supporting steel frame. The cars hang on wheels which are driven by an electric motor operating at 600 volts DC, fed from an extra rail.[10]

The supporting frame and tracks are made out of 486 pillars and bridgework sections. The termini at each end of the line also serve as train depots and reversers.[5]

The current fleet consists of twenty-seven[10] two-car trains[4] built in the 1970s.[5] The cars are 24 metres long and have 4 doors. One carriage can seat 48 with approximately 130 standing passengers.[4] The top speed is 60 km/h and the average speed is 27 km/h.[5]

The *Kaiserwagen* (Emperor's car), the original train used by Emperor Wilhelm II during a test ride on 24 October 1900, is still operated on scheduled excursion services, special occasions and for charter events.[5]

Detail of suspender, wheel and motor

Accidents

15 January 1917

> A train rear-ended another train that had stopped unexpectedly in front of it between Oberbarmen and Wupperfeld, causing the trailing car of the stopped train to fall off of the track. There were two minor injuries.[11] Subsequently, a safety device was developed to make derailments nearly impossible.[5]

21 July 1950

> The Althoff Circus organised a publicity stunt by putting a baby elephant on the floating train at Alter Markt station. As the elephant started to bump around during the ride, she was pushed out the wagon and she fell into the river Wupper.[12] The elephant, two journalists, and one passenger received minor injuries. After this jump, the elephant got the name of Tuffi, meaning 'waterdive' in italian. Both operator and circus director were fined after the incident.

11 September 1968

> A truck crashed into a pillar and caused a section of track to fall. There were no trains in the area at the time. This incident led to the use of concrete walls in pillar anchors.

25 March 1997

Schwebebahn accident of 5 August 2008, near Hammerstein station

Schwebebahn accident of 12 April 1999, near Robert-Daum-Platz station

> A technical malfunction caused a rear-end collision in Oberbarmen station between a structure train and the Kaiserwagen. There were 14 injuries, but no derailment.

12 April 1999

The only fatal accident of the Schwebebahn Wuppertal occurred close to the Robert-Daum-Platz station during maintenance work in the early morning hours of 12 April 1999. Workers forgot to remove a metal claw from the track on completion of scheduled night work. The first train of the day heading east hit the claw at a speed of around 50 km/h, then derailed and crashed down about 10 metres into the river Wupper, killing 5 passengers and leaving 49 injured.[5] The salvage operation took 3 days and nights to complete. 8 weeks after the accident the Schwebebahn went back into operation. The financial damage from the accident was in the vicinity of 8 million Deutsche Mark.

The judicial proceedings following the accident highlighted that the disaster was not caused through technical defects or system failure, but through negligence by workers having fallen behind in their work schedule during the preceding night, and abandoning the work site hastily only 10 minutes before the train departed from the depot. Contributing to the circumstances was a lack of control of their activities by site supervisors.

The Works Manager in charge of safety and the workers dealing with the steel claw at the time were acquitted of all charges by the District Court of Wuppertal. The site supervision personnel, having neglected their duties of control, were sentenced for involuntary manslaughter in 5 cases and bodily injury caused by negligence in 37 cases, but let off on probation with verdict 4 StR 289/01 dated 31 January 2002.[13]

5 August 2008

The Schwebebahn collided with a crane truck making deliveries under the track, causing a 10-metre long tear in the floor of one of the cars. The truck driver was seriously injured, and the train driver and some passengers were treated for shock.[14]

In literature

The Schwebebahn is alluded to in Theodore Herzl's utopian novel *Altneuland* (The Old New Land). For Herzl, the Schwebebahn was the ideal form of urban transport, and he imagined a large monorail built in its style in Haifa.[15]

In film

Rüdiger Vogler and Yella Rottländer use images of the Schwebebahn in Wim Wenders' 1974 movie *Alice in the Cities* (*Alice in den Städten*). It also appears in the 1992 Dutch movie The Sunday Child (*De Zondagsjongen*) by Pieter Verhoeff, in Tom Tykwer's 2000 film *The Princess and the Warrior* (*Der Krieger und die Kaiserin*) and as a background and to a number of outdoor dance choreographies in another Wim Wenders film – 2011's *Pina*, and some dances are set inside the cars.

The Schwebebahn is both subject and title to video work by the Turner Prize-nominated artist Darren Almond. Produced in 1995, *Schwebebahn* is the first of three videos that constitute his Train Trilogy.

In other fiction

Some of the events in Le Feu de Wotan, a Belgian bande dessinée in the *Yoko Tsuno* series, take place in the Schwebebahn. The T.V. series *Thunderbirds* never mentions the name Wuppertal, but often features high-speed trains travelling using suspended rails in a similar style to the Schwebebahn.

See also

- Monorail
- H-Bahn
- Skybus Metro
- Schwebebahn Dresden
- Memphis Suspension Railway

- List of rapid transit systems

References

[1] "Schwebebahn" (http://www.schwebebahn.de/EN/html/fs_start.htm). .

[2] http://www.ag-sdd.de/ausgew_erw/rundgaenge/vollansicht/rundgang_5/spk_9.htm

[3] http://www.wsw-online.de/unternehmen/Download/Geschaeftsberichte/gb2008.pdf Annual report, page 44 (German)

[4] "UrbanRail.Net > Europe > Germany > Wuppertal Schwebebahn (supension railway)" (http://web.archive.org/web/20070219130551/ http://www.urbanrail.de/eu/vrr/wuppertal.htm). Archived from the original (http://www.urbanrail.de/eu/vrr/wuppertal.htm) on 19 February 2007. . Retrieved 20 February 2007.

[5] "Uni Wuppertal – Wuppertal's Suspension Railway: overview and history" (http://web.archive.org/web/20061008161818/http://www. uni-wuppertal.de/wuppertal/schwebebahn/allgemein-en.html). Archived from the original (http://www.uni-wuppertal.de/wuppertal/ schwebebahn/allgemein-en.html) on 8 October 2006. . Retrieved 20 February 2007.

[6] http://www.schwebebahn.de/media/geschichte/flashgeschichte.htm

[7] WZ: Seit 1997 hat der Umbau der Schwebebahn 480Millionen Euro gekostet. 100Millionen Euro mehr als geplant. (http://www. wz-newsline.de/lokales/wuppertal/schwebebahn-ausbau-betrug-bei-millionen-hilfe-1.120928)

[8] WSW-Online: Altgerüst muss verstärkt werden -Schwebebahn außer Betrieb (http://www.wsw-online.de/mobilitaet/Schweben_Sehen/ Ausbau_2010/index.htm)

[9] Railway Magazine, January 2012

[10] "Wuppertaler Stadtwerke AG – English" (http://www.wsw-online.de/schwebebahn/sprachen/english). . Retrieved 20 February 2007.

[11] "http://www.schwebebahn-wtal.de" (http://www.schwebebahn-wtal.de/). . Retrieved 21 February 2007.

[12] Article in Westdeutsche Zeitung, March 2009

[13] http://www.hrr-strafrecht.de/hrr/4/01/4-289-01.php3

[14] http://www.spiegel.de/reise/aktuell/0,1518,570258,00.html

[15] http://yesterdayssalad.wordpress.com/2007/01/03/monorails-were-a-great-idea-in-1902-and-theyre-a-great-idea-now/

External links

- Official Website (http://www.schwebebahn.de/) (German) (English)
- Kaiserwagen Excursion Trip Information (http://new.heimat.de/kaiserwagen/index.php)
- Structurae: Wuppertaler Schwebebahn (http://en.structurae.de/projects/data/index.cfm?id=p00109)
- Wuppertal (http://www.urbanrail.net/eu/vrr/wuppertal.htm) in UrbanRail (http://www.urbanrail.net/)
- The Monorail Society (http://www.monorails.org/tMspages/Wuprtal.html)
- Photo tour of stations new and old (http://mic-ro.com/metro/phototour.html?city=Wuppertal)

DB_Regio

DB Regio AG is a subsidiary of Deutsche Bahn which operates short and medium distance passenger train services in Germany, and operates light and heavy rail infrastructure in the United Kingdom.

Germany

Unlike its long-distance counterpart DB Fernverkehr, DB Regio does not operate trains on its own account. Services are ordered and paid for by the Bundesländer or their Landkreise. Some states have awarded long-term contracts to DB Regio (usually 10 to 15 years).

DB Regio train in Wrocław (Poland)

Sweden

ÖstgötaTrafiken has given DB Regio Sverige a contract to operate local trains for 10 years from December 12, 2010.[1]

UK

Since Deutsche Bahn's purchase of Arriva, DB Regio's operations in the United Kingdom are now known as Arriva UK Trains.

Fleet

Rounded numbers, taken from [2]

- 1,300 locomotives
- 5,300 passenger cars
- 3,100 multiple units

See also

- S-Bahn
- Berlin S-Bahn
- Dresden S-Bahn
- Hamburg S-Bahn
- Hanover S-Bahn
- Leipzig-Halle S-Bahn
- Magdeburg S-Bahn
- Munich S-Bahn
- Nuremberg S-Bahn
- Rhein-Main S-Bahn (Frankfurt / Mainz / Wiesbaden)
- Rhein-Neckar S-Bahn (Ludwigshafen / Mannheim / Heidelberg / Karlsruhe)
- Rhein-Ruhr S-Bahn (Ruhr Area / Düsseldorf / Cologne)
- Rostock S-Bahn
- Stuttgart S-Bahn

External links

- Official DB Regio website [3]

References

[1] "DB Regio wins Swedish contract" (http://www.railwaygazette.com/news/single-view/view/10/db-regio-wins-swedish-contract.html).
     Railway Gazette International. 2009-11-20. .
[2] http://www.db.de/site/shared/de/dateianhaenge/infomaterial/regio/regiokompakt2005__07.pdf
[3] http://www.bahn.de/regional/view/bundesweit/index.shtml

Article Sources and Contributors

Wuppertal-Oberbarmen_station *Source*: http://en.wikipedia.org/w/index.php?title=Wuppertal-Oberbarmen_station *Contributors*: Dewritech, Grahamec

Elberfeld–Dortmund_railway *Source*: http://en.wikipedia.org/w/index.php?title=Elberfeld%E2%80%93Dortmund_railway *Contributors*: ArtVandelay13, Axpde, Grahamec

Wuppertal-Oberbarmen–Solingen_railway *Source*: http://en.wikipedia.org/w/index.php?title=Wuppertal-Oberbarmen%E2%80%93Solingen_railway *Contributors*: Chris the speller, Grahamec

Bergisch-Märkische_Railway_Company *Source*: http://en.wikipedia.org/w/index.php?title=Bergisch-M%C3%A4rkische_Railway_Company *Contributors*: Bermicourt, Doco, Grahamec, Jevansen, Mild Bill Hiccup, Peter Horn, Rich Farmbrough, Slambo, ZH2010

Wupper-Express *Source*: http://en.wikipedia.org/w/index.php?title=Wupper-Express *Contributors*: Grahamec, Mackensen, Markussep

Deutsche_Bundesbahn *Source*: http://en.wikipedia.org/w/index.php?title=Deutsche_Bundesbahn *Contributors*: Ae-a, Alaniaris, ArtVandelay13, Bermicourt, Cvieg, Dewritech, DocendoDiscimus, Doco, Duncanbourne, Ekem, FengRail, G-13114, Henning Makholm, Hmains, Inwind, JNZ, JoniFili, Larry Blake, Morn, Mäfä, Nehrams2020, Poli, ProhibitOnions, Robbie69, Schizzlefuzz, Scriberius, SeNeKa, Slambo, Sunshinemind, Tarannon103, Tim!, Yellowute, Z220info, Zimin.V.G., 9 anonymous edits

Rhine-Ruhr_S-Bahn *Source*: http://en.wikipedia.org/w/index.php?title=Rhine-Ruhr_S-Bahn *Contributors*: ALR997, ArtVandelay13, Bhludzin, CComMack, DerBorg, Dneubert, Doco, GK tramrunner, Grahamec, Iothiania, Kilo-Lima, LOS163, Qualle, Thomas5388, WillaMissionary, Woohookitty, ZH2010, 12 anonymous edits

Rhein-Münsterland-Express *Source*: http://en.wikipedia.org/w/index.php?title=Rhein-M%C3%BCnsterland-Express *Contributors*: Grahamec, Markussep

Maas-Wupper-Express *Source*: http://en.wikipedia.org/w/index.php?title=Maas-Wupper-Express *Contributors*: Grahamec

Hagen_Central_Station *Source*: http://en.wikipedia.org/w/index.php?title=Hagen_Central_Station *Contributors*: Arsenikk, ArtVandelay13, D6, Dneubert, Doco, Grahamec, Kauffner, LittleWink, Mackensen, MadGeographer, Severo, ZH2010, 3 anonymous edits

Remscheid_Central_Station *Source*: http://en.wikipedia.org/w/index.php?title=Remscheid_Central_Station *Contributors*: Grahamec, Kauffner, Oranjblud

Mönchengladbach_Central_Station *Source*: http://en.wikipedia.org/w/index.php?title=M%C3%B6nchengladbach_Central_Station *Contributors*: DerBorg, Dneubert, Doco, Grahamec, Kauffner, MadGeographer, Severo, ZH2010, 1 anonymous edits

Deutsche_Bahn *Source*: http://en.wikipedia.org/w/index.php?title=Deutsche_Bahn *Contributors*: 16@r, A bit iffy, Achmelvic, Adolch, Ae-a, Aecis, Agathoclea, Alan ffm, Alarics, ArtVandelay13, Arwel Parry, Axpde, Axt, Bahnthaler, Basreuwer, Bermicourt, Betacommand, Bigjimr, Bobrayner, Bods, Butterfly0fdoom, BuzzWoof, CambridgeBayWeather, Carrolljon, Cassowary, Chris the speller, Cla68, Closedmouth, Colt9033, CrossHouses, Cyrius, DR, Dan Sellers, DataBeaver, Dcandeto, Dealerofsalvation, DerBorg, Dimboukas, DocendoDiscimus, Doco, Dub8lad1, Ekem, Erianna, Fairlyoddparents1234, FengRail, Fkbreitl, Fratrep, Fudoreaper, G-Man, Gaius Cornelius, Gobeirne, Grafen, Grahamec, Grusl, Gzornenplatz, Henning Makholm, Herr Kent, Hohenberg, Hschaefer, Imgaril, Inwind, JaJaWa, Jansch, JeLuF, Jerryseinfeld, Jnallen, John of Reading, Joolz, JoshFarron, KAMiKAZOW, Kaihsu, Karlthegreat, Keatinga, Kelisi, Klippa, Kozuch, LMB, Lachyl, Langohio, Li-sung, Lightmouse, Likelife, LilHelpa, Lockesdonkey, MakeChooChooGoNow, Mattbr, MaxHD, Mctucker, Mervyn, Meste, Mic, Mkweise, Moe Epsilon, Morven, NE2, Nikai, Nopetro, Only, Oranjblud, Patrick, Pdiddyjr, Peter Horn, Peyodp, Phlebas, Plokijnu, Poli, ProhibitOnions, Prumpf, Qualle, RMJG, RainbowCrane, Ralf.schramm, Ramblersen, Random user 8384993, RedZebra, Redrose64, Reflex Reaction, Rjwilmsi, Roepers, S, Sannse, Scion, SeNeKa, SergioGeorgini, Sf67, Shortfatlad, Simonalexander2005, SkylineEvo, Sladen, Starbois, Sunshinemind, Synchronism, TRBlom, Tabletop, Tazmaniacs, ThePlaz, Tim!, Tobibln, Tompw, Trident13, TrufflesTheLamb, Tschild, TubularWorld, Vegaswikian, Waltonkbbl, WhisperToMe, Wik, Wimox, Woohookitty, Yasirniazkhan, 209 anonymous edits

Wuppertal_Central_Station *Source*: http://en.wikipedia.org/w/index.php?title=Wuppertal_Central_Station *Contributors*: ArtVandelay13, Bermicourt, Chris0693, Dneubert, Glacialfox, Grahamec, John of Reading, Kauffner, MadGeographer, Markussep, SchreiberBike, 1 anonymous edits

German_railway_station_categories *Source*: http://en.wikipedia.org/w/index.php?title=German_railway_station_categories *Contributors*: AdjustShift, ArtVandelay13, Bermicourt, Captainguinness, Chris j wood, Deor, Doco, Fcbfcbfcb, Grahamec, Iste Praetor, Jushi, Sebbl2go, Tam0031, Towopedia, Xyboi, ZH2010, Zweifel, 17 anonymous edits

Wuppertal_Schwebebahn *Source*: http://en.wikipedia.org/w/index.php?title=Wuppertal_Schwebebahn *Contributors*: -jkb-, AGToth, Acps110, Airodyssey, AmosWolfe, Astronaut, Axel-schwede, Axeman89, Bakabaka, Bhludzin, Bobblehead, Bobblewik, Chochopk, Chris 73, Coffeinfreak, Cs302b, Dankarran, DerBorg, DjHammers, Dk 1969, Doco, Dr-Victor-von-Doom, Ekem, Emerson7, Ericamick, Fromgermany, Gestumblindi, GregorB, Ground Zero, Ireneshusband, Jonkerz, Jpatokal, KF, Kelisi, LMB, Legend144, Leonard G., Lightmouse, LorenzoB, Mailer diablo, MapsMan, Millbrooky, MrFish, NawlinWiki, Nbarth, Necrothesp, Nino.shoshia, Occur Curve, Ohconfucius, OlavN, Oxyman42, Peter Horn, Pol098, ProhibitOnions, Psychonaut, Rich Farmbrough, Romanista, Rsrikanth05, Saintswithin, SchuminWeb, Scrabbleship, Signalhead, Sladen, Slambo, TBerlin, ThomasPusch, Thryduulf, Thue, Tintinlover123, TinyMark, Twigmoor, Utcursch, Valmi, W-tal-1, Waterstones, Wildsoda, Wknight94, Woohookitty, ZH2010, 103 anonymous edits

DB_Regio *Source*: http://en.wikipedia.org/w/index.php?title=DB_Regio *Contributors*: Ae-a, CComMack, Carabinieri, CrossHouses, Daigaku2051, DocendoDiscimus, Doco, Henning Makholm, JaJaWa, Jcarls1, John of Reading, Julo, Junafani, Karottenreibe, Likelife, Lupinewulf, Myrtone86, NRTurner, Qualle, Quickliger, Sebastian scha., TubularWorld, Welshleprechaun, Wheeltapper, 25 anonymous edits

Image Sources, Licenses and Contributors

Image:Wsb-04-sonnbornerstr.jpg *Source*: http://en.wikipedia.org/w/index.php?title=File:Wsb-04-sonnbornerstr.jpg *License*: unknown *Contributors*: A.Savin

Image:Wuppertaler Schwebebahn Detail Antrieb.jpg *Source*: http://en.wikipedia.org/w/index.php?title=File:Wuppertaler_Schwebebahn_Detail_Antrieb.jpg *License*: unknown *Contributors*: Atamari, Chris 73, Ies

File:Wuppertaler Schwebebahn Unfall 20080805 0015.jpg *Source*: http://en.wikipedia.org/w/index.php?title=File:Wuppertaler_Schwebebahn_Unfall_20080805_0015.jpg *License*: unknown *Contributors*: User:Atamari

File:12. April 1999 - Ganz Wuppertal steht unter Schock.jpg *Source*: http://en.wikipedia.org/w/index.php?title=File:12._April_1999_-_Ganz_Wuppertal_steht_unter_Schock.jpg *License*: unknown *Contributors*: User:Pillboxs

File:Regio-DB.jpg *Source*: http://en.wikipedia.org/w/index.php?title=File:Regio-DB.jpg *License*: unknown *Contributors*: User:Julo

GNU Free Documentation License Version 1.2, November 2002 Copyright (C) 2000,2001,2002 Free Software Foundation, Inc. 59 Temple Place, Suite 330, Boston, MA 02111-1307 USA Everyone is permitted to copy and distribute verbatim copies of this license document, but changing it is not allowed.

0. PREAMBLE

The purpose of this License is to make a manual, textbook, or other functional and useful document "free" in the sense of freedom: to assure everyone the effective freedom to copy and redistribute it, with or without modifying it, either commercially or noncommercially. Secondarily, this License preserves for the author and publisher a way to get credit for their work, while not being considered responsible for modifications made by others. This License is a kind of "copyleft", which means that derivative works of the document must themselves be free in the same sense. It complements the GNU General Public License, which is a copyleft license designed for free software. We have designed this License in order to use it for manuals for free software, because free software needs free documentation: a free program should come with manuals providing the same freedoms that the software does. But this License is not limited to software manuals; it can be used for any textual work, regardless of subject matter or whether it is published as a printed book. We recommend this License principally for works whose purpose is instruction or reference.

1. APPLICABILITY AND DEFINITIONS

This License applies to any manual or other work, in any medium, that contains a notice placed by the copyright holder saying it can be distributed under the terms of this License. Such a notice grants a world-wide, royalty-free license, unlimited in duration, to use that work under the conditions stated herein. The "Document", below, refers to any such manual or work. Any member of the public is a licensee, and is addressed as "you". You accept the license if you copy, modify or distribute the work in a way requiring permission under copyright law. A "Modified Version" of the Document means any work containing the Document or a portion of it, either copied verbatim, or with modifications and/or translated into another language. A "Secondary Section" is a named appendix or a front-matter section of the Document that deals exclusively with the relationship of the publishers or authors of the Document to the Document's overall subject (or to related matters) and contains nothing that could fall directly within that overall subject. (Thus, if the Document is in part a textbook of mathematics, a Secondary Section may not explain any mathematics.) The relationship could be a matter of historical connection with the subject or with related matters, or of legal, commercial, philosophical, ethical or political position regarding them. The "Invariant Sections" are certain Secondary Sections whose titles are designated, as being those of Invariant Sections, in the notice that says that the Document is released under this License. If a section does not fit the above definition of Secondary then it is not allowed to be designated as Invariant. The Document may contain zero Invariant Sections. If the Document does not identify any Invariant Sections then there are none. The "Cover Texts" are certain short passages of text that are listed, as Front-Cover Texts or Back-Cover Texts, in the notice that says that the Document is released under this License. A Front-Cover Text may be at most 5 words, and a Back-Cover Text may be at most 25 words. A "Transparent" copy of the Document means a machine-readable copy, represented in a format whose specification is available to the general public, that is suitable for revising the document straightforwardly with generic text editors or (for images composed of pixels) generic paint programs or (for drawings) some widely available drawing editor, and that is suitable for input to text formatters or for automatic translation to a variety of formats suitable for input to text formatters. A copy made in an otherwise Transparent file format whose markup, or absence of markup, has been arranged to thwart or discourage subsequent modification by readers is not Transparent. An image format is not Transparent if used for any substantial amount of text. A copy that is not "Transparent" is called "Opaque". Examples of suitable formats for Transparent copies include plain ASCII without markup, Texinfo input format, LaTeX input format, SGML or XML using a publicly available DTD, and standard-conforming simple HTML, PostScript or PDF designed for human modification. Examples of transparent image formats include PNG, XCF and JPG. Opaque formats include proprietary formats that can be read and edited only by proprietary word processors, SGML or XML for which the DTD and/or processing tools are not generally available, and the machine-generated HTML, PostScript or PDF produced by some word processors for output purposes only. The "Title Page" means, for a printed book, the title page itself, plus such following pages as are needed to hold, legibly, the material this License requires to appear in the title page. For works in formats which do not have any title page as such, "Title Page" means the text near the most prominent appearance of the work's title, preceding the beginning of the body of the text. A section "Entitled XYZ" means a named subunit of the Document whose title either is precisely XYZ or contains XYZ in parentheses following text that translates XYZ in another language. (Here XYZ stands for a specific section name mentioned below, such as "Acknowledgements", "Dedications", "Endorsements", or "History".) To "Preserve the Title" of such a section when you modify the Document means that it remains a section "Entitled XYZ" according to this definition. The Document may include Warranty Disclaimers next to the notice which states that this License applies to the Document. These Warranty Disclaimers are considered to be included by reference in this License, but only as regards disclaiming warranties: any other implication that these Warranty Disclaimers may have is void and has no effect on the meaning of this License.

2. VERBATIM COPYING

You may copy and distribute the Document in any medium, either commercially or noncommercially, provided that this License, the copyright notices, and the license notice saying this License applies to the Document are reproduced in all copies, and that you add no other conditions whatsoever to those of this License. You may not use technical measures to obstruct or control the reading or further copying of the copies you make or distribute. However, you may accept compensation in exchange for copies. If you distribute a large enough number of copies you must also follow the conditions in section 3. You may also lend copies, under the same conditions stated above, and you may publicly display copies.

3. COPYING IN QUANTITY

If you publish printed copies (or copies in media that commonly have printed covers) of the Document, numbering more than 100, and the Document's license notice requires Cover Texts, you must enclose the copies in covers that carry, clearly and legibly, all these Cover Texts: Front-Cover Texts on the front cover, and Back-Cover Texts on the back cover. Both covers must also clearly and legibly identify you as the publisher of these copies. The front cover must present the full title with all words of the title equally prominent and visible. You may add other material on the covers in addition. Copying with changes limited to the covers, as long as they preserve the title of the Document and satisfy these conditions, can be treated as verbatim copying in other respects. If the required texts for either cover are too voluminous to fit legibly, you should put the first ones listed (as many as fit reasonably) on the actual cover, and continue the rest onto adjacent pages. If you publish or distribute Opaque copies of the Document numbering more than 100, you must either include a machine-readable Transparent copy along with each Opaque copy, or state in or with each Opaque copy a computer-network location from which the general network-using public has access to download using public-standard network protocols a complete Transparent copy of the Document, free of added material. If you use the latter option, you must take reasonably prudent steps, when you begin distribution of Opaque copies in quantity, to ensure that this Transparent copy will remain thus accessible at the stated location until at least one year after the last time you distribute an Opaque copy (directly or through your agents or retailers) of that edition to the public. It is requested, but not required, that you contact the authors of the Document well before redistributing any large number of copies, to give them a chance to provide you with an updated version of the Document.

4. MODIFICATIONS

You may copy and distribute a Modified Version of the Document under the conditions of sections 2 and 3 above, provided that you release the Modified Version under precisely this License, with the Modified Version filling the role of the Document, thus licensing distribution and modification of the Modified Version to whoever possesses a copy of it. In addition, you must do these things in the Modified Version: A. Use in the Title Page (and on the covers, if any) a title distinct from that of the Document, and from those of previous versions (which should, if there were any, be listed in the History section of the Document). You may use the same title as a previous version if the original publisher of that version gives permission. B. List on the Title Page, as authors, one or more persons or entities responsible for authorship of the modifications in the Modified Version, together with at least five of the principal authors of the Document (all of its principal authors, if it has fewer than five), unless they release you from this requirement. C. State on the Title page the name of the publisher of the Modified Version, as the publisher. D. Preserve all the copyright notices of the Document. E. Add an appropriate copyright notice for your modifications adjacent to the other copyright notices. F. Include, immediately after the copyright notices, a license notice giving the public permission to use the Modified Version under the terms of this License, in the form shown in the Addendum below. G. Preserve in that license notice the full lists of Invariant Sections and required Cover Texts given in the Document's license notice. H. Include an unaltered copy of this License. I. Preserve the section Entitled "History", Preserve its Title, and add to it an item stating at least the title, year, new authors, and publisher of the Modified Version as given on the Title Page. If there is no section Entitled "History" in the Document, create one stating the title, year, authors, and publisher of the Document as given on its Title Page, then add an item describing the Modified Version as stated in the previous sentence. J. Preserve the network location, if any, given in the Document for public access to a Transparent copy of the Document, and likewise the network locations given in the Document for previous versions it was based on. These may be placed in the "History" section. You may omit a network location for a work that was published at least four years before the Document itself, or if the original publisher of the version it refers to gives permission. K. For any section Entitled "Acknowledgements" or "Dedications", Preserve the Title of the section, and preserve in the section all the substance and tone of each of the contributor acknowledgements and/or dedications given therein. L. Preserve all the Invariant Sections of the Document, unaltered in their text and in their titles. Section numbers or the equivalent are not considered part of the section titles. M. Delete any section Entitled "Endorsements". Such a section may not be included in the Modified Version. N. Do not retitle any existing section to be Entitled "Endorsements" or to conflict in title with any Invariant Section. O. Preserve any Warranty Disclaimers. If the Modified Version includes new front-matter sections or appendices that qualify as Secondary Sections and contain no material copied from the Document, you may at your option designate some or all of these sections as invariant. To do this, add their titles to the list of Invariant Sections in the Modified Version's license notice. These titles must be distinct from any other section titles. You may add a section Entitled "Endorsements", provided it contains nothing but endorsements of your Modified Version by various parties--for example, statements of peer review or that the text has been approved by an organization as the authoritative definition of a standard. You may add a passage of up to five words as a Front-Cover Text, and a passage of up to 25 words as a Back-Cover Text, to the end of the list of Cover Texts in the Modified Version. Only one passage of Front-Cover Text and one of Back-Cover Text may be added by (or through arrangements made by) any one entity. If the Document already includes a cover text for the same cover, previously added by you or by arrangement made by the same entity you are acting on behalf of, you may not add another; but you may replace the old one, on explicit permission from the previous publisher that added the old one. The author(s) and publisher(s) of the Document do not by this License give permission to use their names for publicity for or to assert or imply endorsement of any Modified Version.

5. COMBINING DOCUMENTS

You may combine the Document with other documents released under this License, under the terms defined in section 4 above for modified versions, provided that you include in the combination all of the Invariant Sections of all of the original documents, unmodified, and list them all as Invariant Sections of your combined work in its license notice, and that you preserve all their Warranty Disclaimers. The combined work need only contain one copy of this License, and multiple identical Invariant Sections may be replaced with a single copy. If there are multiple Invariant Sections with the same name but different contents, make the title of each such section unique by adding at the end of it, in parentheses, the name of the original author or publisher of that section if known, or else a unique number. Make the same adjustment to the section titles in the list of Invariant Sections in the license notice of the combined work. In the combination, you must combine any sections Entitled "History" in the various original documents, forming one section Entitled "History"; likewise combine any sections Entitled "Acknowledgements", and any sections Entitled "Dedications". You must delete all sections Entitled "Endorsements".

6. COLLECTIONS OF DOCUMENTS

You may make a collection consisting of the Document and other documents released under this License, and replace the individual copies of this License in the various documents with a single copy that is included in the collection, provided that you follow the rules of this License for verbatim copying of each of the documents in all other respects. You may extract a single document from such a collection, and distribute it individually under this License, provided you insert a copy of this License into the extracted document, and follow this License in all other respects regarding verbatim copying of that document.

7. AGGREGATION WITH INDEPENDENT WORKS

A compilation of the Document or its derivatives with other separate and independent documents or works, in or on a volume of a storage or distribution medium, is called an "aggregate" if the copyright resulting from the compilation is not used to limit the legal rights of the compilation's users beyond what the individual works permit. When the Document is included in an aggregate, this License does not apply to the other works in the aggregate which are not themselves derivative works of the Document. If the Cover Text requirement of section 3 is applicable to these copies of the Document, then if the Document is less than one half of the entire aggregate, the Document's Cover Texts may be placed on covers that bracket the Document within the aggregate, or the electronic equivalent of covers if the Document is in electronic form. Otherwise they must appear on printed covers that bracket the whole aggregate.

8. TRANSLATION

Translation is considered a kind of modification, so you may distribute translations of the Document under the terms of section 4. Replacing Invariant Sections with translations requires special permission from their copyright holders, but you may include translations of some or all Invariant Sections in addition to the original versions of these Invariant Sections. You may include a translation of this License, and all the license notices in the Document, and any Warranty Disclaimers, provided that you also include the original English version of this License and the original versions of those notices and disclaimers. In case of a disagreement between the translation and the original version of this License or a notice or disclaimer, the original version will prevail. If a section in the Document is Entitled "Acknowledgements", "Dedications", or "History", the requirement (section 4) to Preserve its Title (section 1) will typically require changing the actual title.

9. TERMINATION

You may not copy, modify, sublicense, or distribute the Document except as expressly provided for under this License. Any other attempt to copy, modify, sublicense or distribute the Document is void, and will automatically terminate your rights under this License. However, parties who have received copies, or rights, from you under this License will not have their licenses terminated so long as such parties remain in full compliance.

10. FUTURE REVISIONS OF THIS LICENSE

The Free Software Foundation may publish new, revised versions of the GNU Free Documentation License from time to time. Such new versions will be similar in spirit to the present version, but may differ in detail to address new problems or concerns. See http://www.gnu.org/copyleft/. Each version of the License is given a distinguishing version number. If the Document specifies that a particular numbered version of this License "or any later version" applies to it, you have the option of following the terms and conditions either of that specified version or of any later version that has been published (not as a draft) by the Free Software Foundation. If the Document does not specify a version number of this License, you may choose any version ever published (not as a draft) by the Free Software Foundation. ADDENDUM: How to use this License for your documents To use this License in a document you have written, include a copy of the License in the document and put the following copyright and license notices just after the title page: Copyright (c) YEAR YOUR NAME. Permission is granted to copy, distribute and/or modify this document under the terms of the GNU Free Documentation License, Version 1.2 or any later version published by the Free Software Foundation; with no Invariant Sections, no Front-Cover Texts, and no Back-Cover Texts. A copy of the license is included in the section entitled "GNU Free Documentation License". If you have Invariant Sections, Front-Cover Texts and Back-Cover Texts, replace the "with...Texts." line with this: with the Invariant Sections being LIST THEIR TITLES, with the Front-Cover Texts being LIST, and with the Back-Cover Texts being LIST. If you have Invariant Sections without Cover Texts, or some other combination of the three, merge those two alternatives to suit the situation. If your document contains nontrivial examples of program code, we recommend releasing these examples in parallel under your choice of free software license, such as the GNU General Public License, to permit their use in free software.

Printed by Books on Demand GmbH, Norderstedt / Germany